BIODIVERSITY

NIPA® GENX ELECTRONIC RESOURCES & SOLUTIONS P. LTD.
New Delhi-110 034

About the Author

Prof. Vir Singh, Ph.D., is an Emeritus Professor of Environmental Science at GB Pant University of Agriculture & Technology, India. He has more than three decades experience in teaching, research, extension, project execution, and research supervision. He has also worked as Coordinator of Liberal Education, Director of Communication, Editor-in-Chief of two monthly magazines published by the university, and Nodal Officer of Disaster Management for the Uttarakhand Government, India. He has been educated and trained in many universities and institutes in India and has also worked with many reputed organizations outside India, including International Centre for Integrated Mountain Development (ICIMOD) based in Kathmandu, Nepal, Galilee International Management Institute (GIMI) in Israel, and Friedrich-Schiller University in Germany. He has a passion for environmental science, ecology, agroecology, and eco-philosophy. He is fascinated by Nature's biodiversity.

Prof. Vir Singh has published 54 books, many monographs, a lab manual, and more than 250 research papers, book chapters, and popular articles. In his book *Fertilizing the Universe* (Cambridge Scholars, 2019), he proposes a theory of astrobiological evolution. His textbook on *Environmental Plant Physiology* (Taylor and Francis, CRC Press, 2020) brings to the fore a botanical strategy for a climate-smart planet. He is also a Climate Reality Leader committed to creating awareness about the ongoing climate change and its long-term implications on every walk of life. He is also actively engaged in environmental writing. His articles on vital contemporary issues are being widely published in English and Hindi dailies and e-magazines.

BIODIVERSITY
Concepts, Crises, and Conservation

Vir Singh
Emeritus Professor
Department of Environmental Science
GB Pant University of Agriculture and Technology
Pantnagar-263145, Uttarakhand, India

NIPA® GENX ELECTRONIC RESOURCES & SOLUTIONS P. LTD.
New Delhi-110 034

NIPA® GENX ELECTRONIC RESOURCES & SOLUTIONS P. LTD.

101,103, Vikas Surya Plaza, CU Block
L.S.C.Market, Pitam Pura, New Delhi-110 034
Ph : +91 11 27341616, 27341717, 27341718
E-mail: newindiapublishingagency@gmail.com
Website: www. nipabooks.com

For customer assistance, please contact
Phone: + 91-11-27 34 17 17
Fax: + 91-11-27 34 16 16
E-Mail: feedbacks@nipabooks.com

ISBN: 978-81-19002-35-1

Composed and Designed by NIPA®.

Preface

Biodiversity not just relates to a count of all the living species, intra-species, and all communities prevailing in the biosphere. Biodiversity also does not merely relate to the richness of life. Biodiversity is not just about its economic attributes. Biodiversity is also not just the spice of life. Biodiversity is the very reality of the mysterious journey of Evolution. Biodiversity prevailing in the biosphere is the most wonderful story of nature's Evolution. Biodiversity is the most pristine 'product' of Evolution. By evolving the diversity of various genotypes of a species, Evolution struck sustainability for all species. By ensuring the sustainability of the species, Evolution ensured the sustainability of all communities. Biodiversity is the glory of Evolution.

Can you imagine Evolution described without the biodiversity in its fold? Look at the very story of Evolution, and you will find it enriching the biosphere by constantly adding biodiversity to it. Our world is a wonderful outcome of biodiversity. Biological Evolution continued adding to newer and newer biological species and in due course of the phenomenon, the human species came into existence.

In the vast biosphere, none of the species can survive, exist, and sustain itself without other species. Interdependence, not independence, is the very essence of the survival, existence, and sustenance of the species. The greater the degree of interdependence the higher the degree of sustainability of the species and the ecosystems they inhabit. Interdependence is the essence of all existence and this phenomenon itself emanates from biodiversity. "Responsibility for life at large obliges us to support diversity, for diversity is not only the spice of life, but it is also a precondition of a vibrant and healthy life," says eco-philosopher Henryk Skolimowski, "moreover, a precondition of the meaningfulness and richness of our individual lives."

The book in your hands sings the glory of biodiversity on planet Earth. Written in a perfect professional tone with an added flavor of emerging concepts that highlight the growing need for biodiversity protection, conservation, and enhancement, the book attempts to make huge promises for the students, scholars, researchers, teachers, and development proponents to articulate the subject matter in their own ways. The human world now deeply understands that their socioeconomic needs can be fulfilled on a sustained basis only when biodiversity blossoms with fullness on the Living Planet. “The Earth’s genes, species, and ecosystems are the product of over 3000 million years of evolution, and are the basis for the survival of our own species,” says Jeffrey A. McNeely, a biodiversity scholar.

Planet Earth is caught in a variety of crises, and all crises are rooted in biodiversity crises. Biodiversity erosion, principally at the hands of the human species, is one of the worst ever crises. Global warming eventually leads to climate change as a result of biodiversity erosion. Biodiversity is responsible for all three phenomena of life, viz, N_2 fixation, photosynthesis, and chemosynthesis. Conservation of biodiversity throughout the length and width of the globe will be phenomenal for restoring our dying planet, averting global warming, and mitigating climate change.

Nature’s resilience and sustainability of ecosystems, and socioeconomic systems are the phenomena that emanate from biodiversity. A sustainable future can also be created on a rich foundation of nature’s biodiversity.

Biodiversity: Concepts, Crises, and Conservation offers professional subject content with fresh achievements in this field. Biodiversity in the literature on the subject available so far does not put much emphasis on biodiversity-sustainability links. The present book highlights, examines, and magnificently explains these vital linkages. Thus, the book opens many windows to a new world of knowledge illuminating new ecological paths toward sustainability that green our hope for ushering in a healthy, vibrant, sustainable, and happy future.

I am so happy to add that the matter on various implicit and explicit aspects, perspectives, and issues synthesized and articulated in this book has been possible due to a happy, stimulating, exuberantly creative, and joyful environment in Canada, especially in Oakville and Toronto of the Ontario province created by my wife Gita, daughter Silvi, son Pravesh, son-in-law Nic, and granddaughter Avery.

I hope the book splendidly and ecstatically fulfills the academic promises it absorbs and assimilates in its professionally enriching subject content.

Vir Singh

Contents

Preface v

Abbreviations and Acronyms xiii

1. Biodiversity: Concepts and Values 1

Defining Biodiversity 1
Levels of Biodiversity 3
Genetic Diversity 3
Species Diversity 4
Community and Ecosystem Diversity 5
Gradients of Biodiversity 8
Latitudinal Effects 8
Altitudinal Effects 9
Biogeographical Classification of India 10
Values of Biodiversity 13
Direct Values 14
Indirect Values 14
Source of Food 14
Drugs and Medicine 15
Cultural and Aesthetic Values 16
Existence Value 16
Bequest Value 17
Option Values 17
Ecosystem Services 17
Summary 18
Exercises 20
Multiple Choice Questions 20
Short-Answer Questions 23
Long-Answer Questions 23

References 24
Web Resources 24

2. Agrobiodiversity: Biodiversity in Agriculture........25

Biodiversity, Survival, and Sustenance 25
Biodiversity and Agrobiodiversity 26
Biodiversity through the History of Agriculture Development 27
The Four Eras of Agriculture 27
Biodiversity and Food Production Scenario 31
Pyramid of Agriculture 34
Agriculture and A Sustainable Future 37
Agriculture Pyramid and Fate of Humanity 38
Mountain Agroecosystem Specificities 40
Mainstream Agriculture vs Mountain Agriculture 41
Agroecosystem Diversity 42
Uncultivated Land 45
Cultivated Land 47
Livestock 47
Households 49
Wild Relatives of Crop Plants in the Himalayan Region 50
Summary 52
Exercises 54
Multiple Choice Questions 54
Short-Answer Questions 57
Long-Answer Questions 57
References 58
Web Resources 58

3. Biodiversity of the Pedosphere 59

The Pedosphere 59
Pedodiversity 60
An “Invisible World” of Biodiversity 61
Soil Flora 64
Microflora 64
Macroflora 66
Soil Fauna 67

Microfauna 67
Mesofauna 67
Macrofauna 67
Megafuana 68
Soil Biodiversity Erosion 69
Conserving Soil Biodiversity 71
Summary 72
Exercises 74
Multiple Choice Questions 74
Short-Answer Questions 77
Long-Answer Questions 77
References 77
Web Resources 78

4. Biodiversity in the Chemosynthesis-Based Community 79

The Two Types of the Living Communities 79
The Photosynthesis-Based Communities 80
The Chemosynthesis-Based Community 81
Biodiversity in Hydrothermal Vent Ecosystems 83
Threats to the Chemosynthesis-Based Biodiversity 86
International Obligations 87
Summary 88
Exercises 90
Multiple Choice Questions 90
Short-Answer Questions 93
Long-Answer Questions 93
References 94
Web Resources 95

5. Threats to Biodiversity 97

Destruction of Habitats 97
Habitats Fragmentation 98
Introduction of Exotic Species 98
Hunting and Overexploitation 100
Poaching and Smuggling 100
Agriculture 100

Environmental Pollution 101
Man-Wildlife Conflicts 101
Measures to Avert Man-Wildlife Conflicts 102
Extinction of Species 103
Causes of Extinction 103
IUCN Red List 106
Summary 107
Exercises 109
Multiple Choice Questions 109
Short-Answer Questions 113
Long-Answer Questions 113
References 113
Web Resources 114

6. Biodiversity Conservation 115

Strategies for Biodiversity Conservation 116
In-Situ Conservation Strategies 116
Ex-Situ Conservation Strategies 131
Hotspots of Biodiversity 134
Megadiverse Countries 135
International Efforts for Biodiversity Conservation 137
Summary 138
Exercises 140
Multiple Choice Questions 140
Short-Answer Questions 144
Long-Answer Questions 144
References 145
Web Resources 145

7. Biodiversity and Sustainability 147

Ecological Sustainability: A Vital Dimension of Sustainability 147
Biodiversity-Sustainability Links 148
Biodiversity, Economic Development, and Sustainability 149
Natural vs. Anthropogenic Ecosystems 151
Biodiversity-Based Sustainability Principles 152

Soil Biodiversity 152
Soil Surface Biodiversity 153
Nutrient Cycles 154
Sustainability Management... 155
Summary 157
Exercises 159
Multiple Choice Questions.. 159
Short-Answer Questions 162
Long-Answer Questions 163
References.. 163
Web Resources... 163

Glossary 165

Subject Index...171

Abbreviations and Acronyms

ATP	adenosine triphosphate
BNF	biological nitrogen fixation
Bt	*Bacillus thuringiensis*
C	carbon
$C_{19}H_{22}O_6$	gibberellic acid
$C_6H_{12}O_6$	glucose
CBD	Convention on Biological Diversity
CH_4	methane
cm	centimeter
cm^2	square centimeter
cm^3	cubic centimeter
CO	carbon monoxide
CO_2	carbon dioxide
CPR	common property resource
CR	critically endangered
DD	data deficient
DNA	deoxyribonucleic acid
EN	endangered
EU	European Union
EW	extinct in the wild
EX	extinct
FAO	Food and Agriculture Organization
Fe	iron
g	gram
GA	gibberellic acid
GMO	genetically modified organisms
GPP	gross primary productivity
H_2	hydrogen
H_2O	water (hydrogen monoxide)
H_2S	hydrogen sulfide

H_2SO_4	sulfuric acid
HCN	hydrogen cyanide
HRVs	high responsive varieties
HYVs	high yielding varieties
IPCC	Intergovernmental Panel on Climate Change
IUCN	International Union for Conservation of Nature and Natural Resources
km	kilometer
km^2	square kilometer
LPG	liberalization, privatization, and globalization
m	meter
m^2	square meter
m^3	cubic meter
MAB	Man and Biosphere
mm	millimetres
NO_2^-	nitrites
NO_3^-	nitrates
°C	degrees Celsius
SO_4^{2-}	sulfates
SOC	soil organic carbon
SOM	soil organic matter
UK	United Kingdom
UN	United Nations
UNEP	United Nations Environment Programme
UNESCO	United Nations Educational, Social and Cultural Organization
USA	United States of America
USAID	United States Agency for International Development
VU	vulnerable
WCED	World Commission on Environment and Development
WDPA	World Database on Protected Areas
WMO	World Meteorological Organization
WWF	World Wildlife Fund for Nature

1

Biodiversity Concepts and Values

The history of evolution reveals that life on Earth has gradually gone enriching itself by adding more and more varieties to it. A variety of organisms as of today exhilaratingly blossoms in the biosphere. The organisms are divided into three broad categories based on their distinctive characteristics, viz. plants, animals, and microorganisms. The other broad categories all the organisms on Earth are placed in are the five kingdoms, viz. Protista, Monera, Fungi, Plantae and Animalia. These kingdoms encompass living organisms based on their cell type (prokaryotic or eukaryotic), cell number (unicellular or multicellular), genetic recombination type (binary fission, asexual or sexual type), or feeding behavior or mode of nourishment (autotrophic or heterotrophic).

The kingdoms are further divided into various harmonized groups. These include division (among plants), phyla (among animals), classes, orders, families, genus, and species. There is enormous biodiversity at every level in each of the kingdoms.

Every species in every kingdom of the biosphere, in essence, is unique in itself. This uniqueness of organisms constitutes the basis of biodiversity. These organisms in their multiple numbers (populations) organize in their specific communities. Each community is unique in itself, in terms of the variety of the species and operating environmental factors, and in terms of its structure and functions. In every community, again there are many kinds of genotypes – specific genetic characteristics – of all species of plants, animals, and microorganisms. Biodiversity is the greatest wonder on Earth and the very basis of ecological integrity and the sustainability of life on Earth.

Defining Biodiversity

The term biological diversity was coined in 1980 by Thomas Lovejoy in the book Conservation Biology (Soulé and Wilcox 1980). The contracted term "Biodiversity" was coined by W. G. Rosen in 1985 (Wilson 1988). The field of biodiversity, with interdisciplinary relationships, has expanded at a very rapid pace in just three decades (see, for example, Wilson 1980, 1992; Shiva et al. 2005; McNeely et al. 2006; Nautiyal et al. 2012).

Biodiversity, or biological diversity, means the variability among living organisms from all sources, for example, all terrestrial and aquatic ecosystems. It includes varieties of all organisms as well as the complexes they create. There is enormous biodiversity on Earth which itself is a wonderful telltale of natural evolution. Biodiversity, in essence, is the total of all genes, all species, and all ecosystems of the biosphere. The beauty of the Earth is owing to its biodiversity (Fig. 1.1)

Fig. 1.1: The beauty of our Living Planet is attributable to its biodiversity (image by Alain Audet from Pixabay)

Not all species prevail in the same place. Environmental factors operating on a site and the range of tolerance of a species determine the occurrence of a species. It is why biodiversity varies from region to region, area to area, and place to place in the same area. Thus, taking into consideration all the habitats of plants, animals, and microorganisms, we can imagine the enormous count and extent of biodiversity in the Earth's biosphere.

The human species depends on nature's biodiversity for various purposes and is itself an indistinguishable part of the biodiversity. Biodiversity is a precondition for the survival and materialistic progress of human populations. There also exists amazing diversity among human beings at the genetic level. Cultural variability in humankind is also a manifestation of nature's biodiversity. Spurt in human populations and their rate of expansion following the destruction of natural habitats and insatiable greed have become a major cause of biodiversity depletion, destruction, and pushing species towards extinction.

Levels of Biodiversity

Life and life's systems are invariably complex. There exists a wide variety of living species with amazing genetic variations among their individuals. Again, the genetically distinctive individuals of every species living in a variety of ecological systems amidst complex ecological relationships. An understanding of biodiversity is crucial for the recognition of its multiple values and its conservation. Under the "umbrella' of the biosphere, biodiversity prevails on three hierarchal levels: i) genetic diversity, ii) species diversity, and iii) community and ecosystem diversity. With these three distinctive characteristics, biodiversity prevails amidst the state of ecological balance and ecological integrity.

Genetic Diversity

Every individual species is a rich repository of genetic information. The number of genes in a species also varies from that in others. For example, the human species (*Homo sapiens*) has as many as 35000 to 45000 genes to display (20500 genes, according to an estimate), rice (*Oryza sativa*) 32000 to 50000 genes, fruit fly (*Drosophila melanogaster*) 13000 genes, and colon bacteria (*Escherichia coli*) 4000 genes. Genetic variation among all species is due to these varying numbers of genes or due to varying genetic information. The varying genetic information determines certain specific characteristics of a species.

The genetic variation within a species is referred to as genetic diversity. The genetic variation may be in chromosomal structures, in entire genes, i.e. the traits determining particular characteristics, or in alleles, i.e. different variants of the same genes. Genetic diversity is vital for the population of a species to adapt to its environment and respond to natural selection.

The higher the measure of genetic diversity the better the ability of a species to adapt to changed environmental conditions. A narrow base of genetic diversity results in uniformity. Genetically similar crops give rise to monocultures. A trend of raising crop monocultures is encountered in modern agriculture. Monocultures may provide higher yields of food grains; however, they are ecologically vulnerable and are liable to be destroyed due to onslaughts of various insects and pathogens. A species with a broader genetic base, that is, with a high degree of genetic diversity, has a high degree of resilience and is ecologically more stable.

Genetic variability within a species, or intra-species diversity, often registers an increase in environmental variability. The evolution of new species – the speciation – depends upon the amount of genetic variability. Speciation is vital in maintaining a high degree of diversity at species and community levels. The higher the number of species in a community, the higher the level of genetic diversity it harbors, and vice versa.

Species Diversity

A variety of species in a region refers to species diversity. All the species of plants, animals, and microorganisms, or across all the five living kingdoms in the biosphere count as species diversity. The species are the most distinctive units of biodiversity and have a specific role to play in an ecosystem. The loss of a species has an impact on the structure and functioning of an ecosystem.

We use two basic parameters to determine the species diversity of a site or region, viz. i) species richness, and ii) evenness or equitability.

Species richness denotes the number of species per unit area. The species number increases with an increase in the area of the site. Generally, the higher the species richness the greater the species diversity.

Evenness or equitability denotes the evenness in the number of individuals of a species. The evenness of a species is representative of higher species diversity in a region. Consider the following examples to understand the species evenness and species diversity relationship:

- Sample Area A: Three species of a butterfly, out of which two species are represented by one individual each; the third one has four individuals.
- Sample Area B: Three same species of butterfly each represented by two individuals.
- Sample Area C: Three species of butterfly, a bird, and a rabbit

In the above-cited examples, it can be inferred that:

- Sample Area B shows greater evenness than Sample Area A with equal chances for the representative species to occur, which reveals greater diversity in Area B as compared to Area A.
- Sample Area C has taxonomically unrelated species, thus harboring the highest species diversity among all the three sample areas.

- Each sample area depicts an equal number of species, however, different numbers of individuals per species. Thus, the number of species, as well as several individuals per species, determine the measure of species diversity in an area.

The ecological state of a species being unique to a defined geographic location is referred to as endemism. Organisms indigenous to a place would not be counted as endemic if found elsewhere. The extreme opposite of endemism denotes cosmopolitan distribution.

Community and Ecosystem Diversity

Species organize in populations and populations in communities. All biotic communities are not alike. They reveal enormous variations among them.

Community diversity reflects variations in the biotic communities in which the species thrive. There are three perspectives relating to community diversity (Fig. 1.2): i) alpha diversity, ii) beta diversity, and iii) gamma diversity.

Alpha Diversity

Alpha diversity means diversity within a community, that is, the diversity of the organisms sharing a common community. Both species richness and evenness or equitability are applied to measure diversity within a community.

Beta Diversity

Beta diversity refers to diversity between communities. Communities are not always alike. They change and with them, their species also change. Not two or more communities have the same species. Species composition changes due to certain environmental gradients, such as altitudinal, temperature, and moisture gradients. The higher the heterogeneity in the communities of a region the greater the measure of beta diversity between them.

Fig. 1.2: Alpha, beta, and gamma diversity: the three aspects of biodiversity

Gamma Diversity

Gamma diversity means the diversity of the communities or habitats over a geographical area or the entire landscape. It generally reflects the species turnover rate concerning the distance between sites of similar habitats or with expanding geographical areas. The gamma diversity applies to larger geographical areas. A high level of diversity is the basis of higher productivity and stability of a community. A more diverse community is functionally more healthy, vibrant and sustainable (Fig. 1.2)

A community is a biotic content of an ecosystem. An ecosystem comprises both biotic and abiotic components in reciprocal interaction with each other. Therefore, ecosystem diversity is distinguishable from community diversity in terms of the variations in the structure and functions of an ecosystem.

The ecosystem diversity describes a variety of niches, trophic levels, and ecological processes sustaining the food chains, food webs, and nutrient cycles. The factors that affect ecosystem diversity are shown in Fig. 1.3.

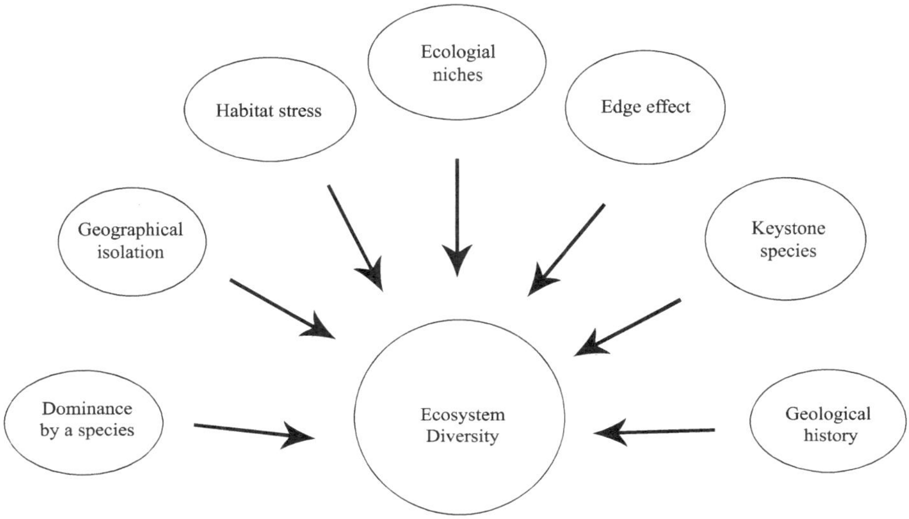

Fig. 1.3: Factors affecting the degree of an ecosystem's diversity

A complex ecosystem provides a large number of ecological niches offering an appropriate range for the activities of the organisms within a community. The larger the number of ecological niches, the greater the degree of ecosystem diversity. If an area gets isolated due to certain meteorological or geological factors, it is generally found harboring a limited number of species. For example, if an island loses its species due to a tsunami, it would be difficult for it to regain that species from another area. A dominant species in an ecosystem tends to monopolize natural resources at the expense of others, thus imposing some restrictions on an increment in the biodiversity of an ecosystem. Ecotones or transition areas show a comparatively higher degree of biodiversity, which is an edge effect. Keystone species determine the ability of a large number of other species to persist in the community, thus playing a crucial role in ameliorating the environment for many other species to add to ecosystem diversity. All top carnivores, e.g. lion, tiger, crocodile, etc. are considered keystone species as they are the ones indirectly regulating the populations of other animal species. If a keystone species vanishes, it would trigger the processes of ecological degradation.

Geological history also happens to be a decisive global factor enhancing or restricting or curbing biodiversity in a region. For example, arctic environments did not allow most of the species to flourish, but, on the other hand, well-established ecosystems such as rainforests allow a very high degree of biodiversity to blossom in their environments. Habitat stress due to biotic or abiotic factors creates conditions unfavorable for the biodiversity to blossom in an area.

There are some other interesting terms concerning community and ecosystem diversity. Point diversity refers to diversity at the smallest scale. In other words, point diversity denotes the diversity of a microhabitat. Epsilon diversity, on the other hand, refers to the diversity of a group of areas of gamma diversities and is also known as regional diversity.

Gradients of Biodiversity

Biodiversity patterns do not prevail uniformly across the globe. Apart from some local natural and anthropogenic factors, the two dominant factors that markedly affect biodiversity are latitude and altitude. Variations in the natural biodiversity occur following the changes in latitude and altitude. Biodiversity gradually increases from high to low latitude, i.e., from the poles to the equator of the Earth.

Latitudinal Effects

The climate in the temperate regions of the Earth has not been stable over long geological times. In these regions towards the glaciated poles of the Earth, the climate is more severe and quite inhospitable for life. Species that evolved in this region tended to migrate toward favorable environments. Further, the annual period for the growth of the plants is short. The photosynthetic base for the support of consumer species is quite narrow. The pressure of pests, parasites, and disease-causing organisms is not up to mark in cold climates. Therefore, only a few dominant species prevail in temperate environments. The lowest level of biodiversity is encountered in the cryosphere where only a few species of animals can survive.

Biodiversity prevails in high magnitude and abundance in the tropics towards lower latitudes, where ecosystems are more complex, heterogeneous, and stable. The high level of biodiversity we encounter in the tropics is attributable to some of the general characteristics of these geographical areas (Anubhav et al. 2009, Singh 2009, Ormsby and Bhagwat 2010) as pointed out below.

- The tropical areas of the planet have not been glaciated. The climate of the tropical region has been quite stable and so do the natural ecosystems over long geological times, thus making the habitats more favorable for the continuous evolution and prevalence of species.

- Tropical communities have had pretty long geological times for evolution with a pretty high degree of speciation and favorable conditions for adaptation and sustenance.
- Tropical areas receive more solar radiation which is vital for creating a healthy, vibrant and stable base of the producers (the photosynthesizers) providing energy and nutrients for a large number of organisms and thus supporting a wider range of species and their genotypes.
- Tropical areas receive a very high amount of precipitation which is vital for the forest ecosystems dominated by trees and evergreen vegetation that in turn create an environment for feeding, supporting, and sustaining a large number of species by creating more trophic levels, food chains, and complex food webs.
- Environments with warm temperatures and high humidity with a longer duration of high-intensity solar radiation are more favorable for photosynthesizers, the producers, for supporting communities with a variety of species of plants, animals, and microorganisms and thus playing a crucial role in the conservation of biodiversity.
- A diverse community provides appropriate conditions for a variety of pests, parasites, and disease-causing organisms that help maintain an ecological balance by not letting a single or a few species to prosper at the cost of biodiversity.
- Tropical communities make favorable environments for the natural development of ecological niches for a variety of specialist species thus bringing more enhancement to life blossoming within natural ecosystems.
- Higher outcrossing rates among organisms, especially among plants, in the tropics, are conducive to building up and maintaining higher levels of genetic variability.
- Higher amounts of soil organic matter (SOM) and soil organic carbon (SOC) as a result of soil-plant relationships help maintain soil fertility level that, in turn, helps build up a healthy and vibrant soil ecosystem supporting enormous biodiversity within the soils (Singh 2019, 2020).

Altitudinal Effects

Biodiversity decreases from lower to higher altitudes in the mountains. As we ascend 1000 meters, we experience a decrease in temperature to the tune of about 6.5^{0}C. Thus, the temperature in the mountain ecosystems goes on decreasing along an increase in altitude with subsequent increase in seasonal variability making the environments less favorable for a variety of species. Edaphic factors influenced by high rates of erosion and catastrophic landslides – like shallow soil, a high proportion of gravel in the soil, emergence of rocks, etc. – also affect the level of biodiversity in larger areas in the mountain ecosystems. Biodiversity significantly decreases in the rare habitats at very high altitudes beyond timberline in alpine climate in the mountains, for example in the Himalayas. Glaciers of the mountains are witnessed with the lowest level of species diversity.

Biogeographical Classification of India

The biosphere of the Earth is stocked with amazing biodiversity. Some of the geographic regions of the planet are especially rich in biodiversity. India – a megadiverse country – is one of them. Of the total geographical area of the globe, India occupies only 2.4%. However, India contributes as much as 8% of its known biodiversity to the world's known biodiversity. In India too, not all the areas are uniform in terms of their diversity. Some of the areas are richer in biodiversity than others while some are rarely endowed with biodiversity. Let us first have a glimpse of the 10 biogeographical zones India is categorized into. Varying environmental conditions and species groupings are the basis of the biogeographical categorization of India. Each of the zones has its characteristic climate, soil, water sources, and topographical features and gives refuge to distinctive flora and fauna. The general features of these biogeographical zones are described in Table 1.1 (see Fig. 1.4 also).

Table 1.1: India's biogeographical zones and their specificities

Biogeographical zone	Location of the zone	Area, %	Major species
1. Trans-Himalayan Zone	High mountains and cold deserts of Ladakh, Jammu and Kashmir, Lahul and Spiti areas of Himachal Pradesh, North Sikkim.	5.6	Sparse alpine steppe vegetation with several native species of herbs and grasses; wild goats and sheep, snow leopard, migratory blacknecked crane (*Grus nigricollis*)
2. Himalayan Zone	Himalayan mountain region comprising the Indian states of Jammu and Kashmir, Himachal Pradesh, Uttarakhand, Sikkim and Arunachal Pradesh and Union Territory of Ladakh and the Darjeeling district of West Bengal	6.4	Vegetation in alpine and forest ecosystems embraces several native species of trees, shrubs and herbaceous plants; Rare and endangered species like Musk deer (*Moschus moschiferus*), Hangul (*Cervus eldi eldi*), Himalayan Tahr (*Hermitragus jemlabicus*) and Takin (*Budoreas taxicolor*), and many bovid species, like Markhor (*Capra falconeri*), Bharal (*Pseudois nayaur*) and Ibex (*Capra ibex*)
3. Desert Zone	The Desert of Thar and Kutch in Rajasthan and Gujarat	6.6	Large expanse of grasslands with many species of grasses and other herbaceous plants interspersed with shrubby species; desert cat (*Felis libyca*), wolf (*Canis lupus*), and bird species like the Great Indian Bustard (*Ardeotis nigriceps*) and Houbara Bustard (*Chamydotis undulate*)
4. Semi-Arid Zone	Transition zone between deserts and dense forests of Western Ghats	16.6	Several species of grasses, herbs and shrubs; Canine species such as the Lion (*Leo persica*), an endangered species confined to a small area in Gujarat; Wolf (*Canis lupus*), Caracal (*Felis caracal*) and Jackal (*Canis aureus*); and the cervid species such as Chital (*Axis axis*) and Sambar (*Cervus unicolor*)
5. Western Ghats Zone	The mountains along the west coast of Indian peninsula extending from the peninsula's southern tip (8^0N) northwards about 1600 km to the mouth of the river Tapi (21^0N), one of the world's unique biological regions	4	Evergreen tropical forests with many endemic species; The Cane turtle (*Heosemys silvatica*) and the Travancore Tortoise (*Indotestudo forstem*), the two endangered taxa confined to a small area in central Western Ghats; Malabar Grey Hornbill (*Ocycerous griseus*), Malabar Civet (*Viverricula megaspila*), Nilgiri Langur (*Presbytis jobni*), Nilgiri Tahr (*Hemitragus bylocrius*), Lion Tailed Macaque (*Macaca silenus*), and Grizzled Giant squirrel (*Ratufa macroura*).

Biogeographical zone	Location of the zone	Area, %	Major species
6. Deccan Peninsula Zone	A semi-arid region falling in the Western Ghats rain shadow area, the largest biogeographical zone of India covering forest areas particularly in parts of Madhya Pradesh, Maharashtra and Odisha.	42	Deciduous forests, thorn forests and degraded scrubland; Wild buffalo (*Bubalus bubalis*), Swamp Deer (*Cervus duvauceli*), Nilgai (*Boselapbus tragocamelus*), Chousingha (*Tetracerus quadricornis*), Sambhar (*Cervus unicolor*) and Chital (*Axis axis*)
7. The Gangetic Plains	Topographically by and large homogeneous	10.8	Swamp Deer (*Cervus duvauceli*), Hog-Deer (*Axis porcinus*), Hispid Hare (*Carpolagus bispidus*), Buffalo (*Bubalus bubalis*), Elephant (*Elephas maximus*) and Rhino (*Rhinoceros unicornis*)
8. The North-East India	The transition zone between the Indian, Indo-Malayan and Indo-Chinese biogeographical regions; a meeting point of the Himalayan mountains and Peninsular India, A biodiversity hot spot (Eastern Himalaya) in India; a gateway for most of the flora and fauna of India	5.2	Many of the species are confined to this zone, some are found in some areas of the Khasi Hills
9. The Indian Islands	A group of 325 islands, 21 inhabited islands, Andaman to the north and Nicobar to the south	0.3	Tropical moist evergreen forests with flora and fauna not found elsewhere; very high degree of endemism; they support a wide variety of corals; the only representative of endemic island biodiversity; Narcondam Hornbill and South Andaman krait are some species occurring in this zone
10. The Coastal Region	The 5423 km long coastline from Gujarat to Sundarbans; mangroves, sandy beaches, mud flats, coral reefs and marine angiosperm pastures	2.5	As many as 25 islets of coral origin rich in biodiversity with a typical reef lagoon system constitute Lakshadweep

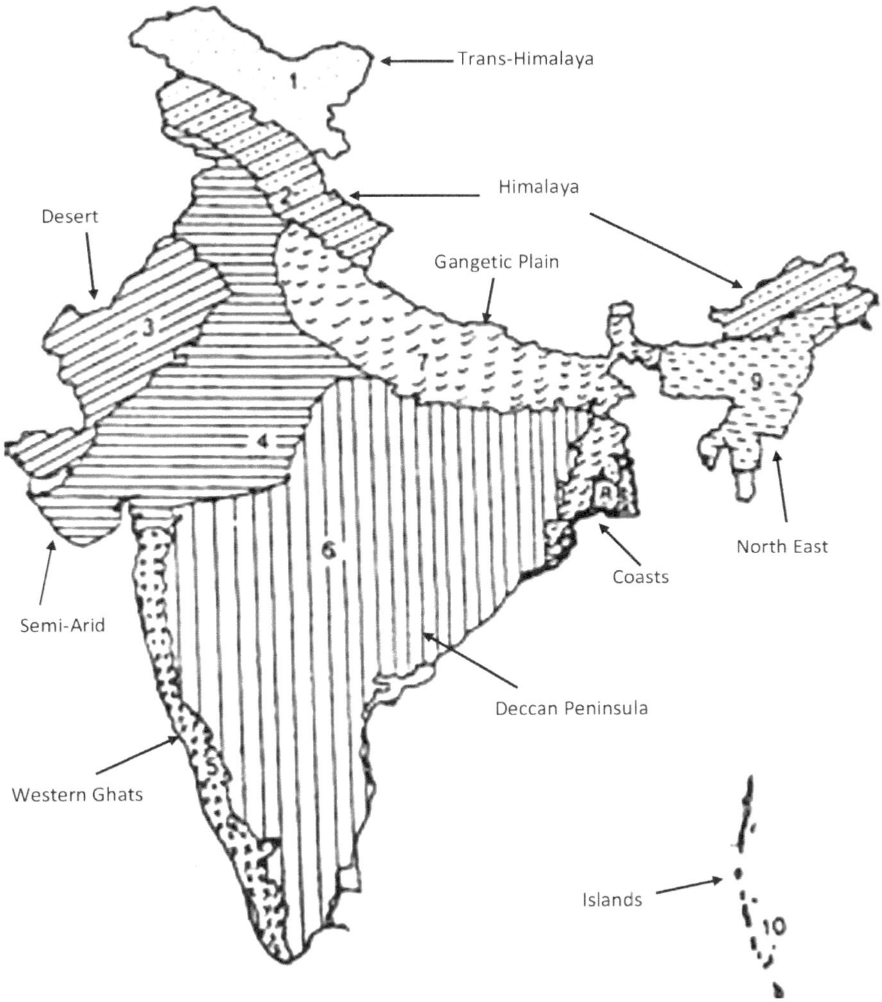

Fig. 1.4: Biogeographical zones of India

Values of Biodiversity

The genes, the species, and the ecosystems are the basis of biological resources that lay the foundation of life on Earth. Species are the building blocks of ecosystems and ecosystems provide the life-support systems for all living species. When we ponder over biological resources from a human perspective, we happen to elaborate on different values on a scale from ecocentric to anthropocentric. It would be ethical that ethical values of biodiversity be a starting point for an appropriate and sound attitude of humans and their institutions towards nature.

The World Charter for Nature adopted by the General Assembly of the United Nations in 1982 recognizes that humankind is a part of nature, that every form of life is unique and it warrants respect towards nature regardless of its (utilitarian) worth to human beings.

Anyhow, certain classifications of values of biological resources have been developed such as (McNeely 1990):

a. Direct Values

(1) Consumptive Use Value (non-market value of firewood, fodder, medicine, game, etc.)

(2) Productive Use Value (commercial value of timber, fish, crops, etc.)

b. Indirect Values

(1) Non-consumptive Use Value (scientific research, recreation, etc.)

(2) Option Value (value of maintaining options available for the future)

(3) Existence Value (value of the ethical feeling of the existence of nature, etc.)

In this classification, we miss – as a direct or indirect value – the Life Support Value, which corresponds to the maintenance of ecological processes.

The diversity in biological resources – genes, species, and ecosystems – provides the basis with which different human conditions can adapt to changes, and the basis for many different social, and cultural expressions as well (Shiva et al. 2005).

Source of Foods

All foods of all heterotrophs come from living biological resources. We derive a variety of foods – fruits, pods, seeds, edible flowers, buds, leaves, stems, roots, mushrooms, etc. – from the plants. Honey we find to consume is made up of a variety of flowering plants. Milk, one of the widely consumed animal products, also comes from a variety of plants used as fodder by milch animals. The foods processed in our kitchens involve a variety of plants. A variety of dishes people are so fond of includes edible products from several plants. Fermented foods, curd, buttermilk, etc. are prepared with the help of several microorganisms. Alcoholic drinks and beverages available in the market also involve the use of a variety of microorganisms in the process of their preparation.

Nearly 85 percent of all foods are produced employing cultivated plants. As much as about 75% of the total energy humanity depends upon is derived from just three carbohydrate-rich food grain crops belonging to the Poaceae or grass family. These are wheat, rice, and corn. However, numerous varieties (genotypes) of these food grain crops are there being cultivated and utilized by human populations throughout the world. Among them are varieties of landraces and those developed using breeding principles. The latter are referred to as high-yielding or improved varieties. Plant breeders have also developed improved varieties of pulses, fruits, vegetables, oilseeds, sugarcane, and of all other food crops. The average life of a crop variety is often short, say 5 to 15 years. After the lapse of this period, it begins to deteriorate. Therefore, the regular development of new varieties becomes a necessity. Genes of the landraces and/ or wild relatives of the food crops are vital for developing varieties laden with the traits like drought and pest resistance.

Nearly 15% of human foods are produced by animals. New breeds of animals using genetic material from native animals have been developed for obtaining milk, eggs, wool, meat, etc. in higher quantities. In our world, fascinating breeds of all types of domesticated animals – cattle, buffalo, sheep, goats, horses, mules, poultry, etc. – are put on various services to humankind. Not only are the animal breeds serve as sources of food and fiber but also for various agricultural operations and carting of agricultural produce in rural areas.

Drugs and Medicines

People have been using various plants for the treatment of various ailments. *Jadi-booti* is a common notion being used for the selected medicinal plants being extensively utilized for the prevention and treatment of different ailments. The science of Ayurveda is replete with knowledge about a variety of trees, shrubs, and herbs of specific medicinal properties used in treating specific diseases. In our modern world, several pharmaceutical companies rely upon several plants with therapeutic values.

The diversity of the plants providing substances and active principles that are developed into valuable drugs exists in diverse habitats. Quinine, a very popular drug used for the treatment of malarial parasites is obtained from *Chincona ledgeriana*. Morphine used as an analgesic drug is obtained from *Papaver somniferum*. The anti-cancer drug Taxol is derived from the bark of the yew tree, *Taxus baccata*. Most of the traditional drugs in the world are prepared from a variety of plants. Many of the valuable plants grow in the wilderness, and many are being cultivated.

Cultural and Aesthetic Values

Biodiversity has a great place in our religious and cultural life. Many of the plants and animals are regarded as symbols of national pride and cultural heritage. Pipal (*Ficus religiosa*) is worshipped by the Hindus. Regarded as a sacred plant, Tulsi (*Ocimum sanctum*) is grown and worshipped in devout Hindu families. Some wild animals, domesticated animals, birds, and even snakes are worshipped on special occasions. We have specific national and provincial trees, flowers, and animals considered sacred and are given legal protection.

Biodiversity has numerous aesthetic rewards for us. Landscapes harboring a greater degree of biodiversity are laden with natural beauty. Walking amidst the beauty of nature brings a lot of joy and creative feelings. We love to visit places with scenic beauty because beauty is a need of the human soul. Without beauty in life, we shrink. Eco-tourism is a new trend in the tourism industry. Visit wildlife sanctuaries, national parks, biosphere reserves, and places with aesthetic appeal to allure people as we have an inner quest to experience more and more life – that is, more and more biodiversity – blossoming around.

Existence Value

Existence value means the value of ethical feelings about the existence of nature. We often attach value to many species even if we know that we would never encounter them. We appreciate their existence and get fascinated by the very feeling that they exist in a particular environment or habitat.

All species or genetic resources as an integral part of biodiversity are not of direct human use. It, however, does not imply that they are valueless to us. They too have value for their existence on Earth. Their existence value often reflects in the sense of well-being. For example, we know that biodiversity prospering around hydrothermal vents on ocean floors is hard of any direct use and there is hardly any desire to visit it. However, we derive benefit in terms of the knowledge of its existence and are fascinated by the very feeling of its existence. Till 1977 we knew that most marine life is dependent on photosynthesis. But with the discovery of hydrothermal vents, especially of the giant tube worm, we came to know about the chemosynthetic community of the hydrothermal vents which is attributable to the existence value of this wonderful biodiversity contributing to the enrichment of our knowledge about biodiversity.

Bequest Value

A natural resource as such or a component of it or its biodiversity or ecosystem functioning may not be of any concern or use directly or indirectly. However, the present generation values it, ensuring its availability to future generations. This is what is called bequest value. This emanates from a person's concern that our future generations should have access to natural resources and opportunities built around them. In other words, the non-use value of biodiversity or ecosystem function in our present times is perceived to be of use value for future generations. This indicates a perception of valuing biodiversity from the current knowledge that the resources and opportunities based on them are being passed on to descendants.

Option Values

There may be several species, especially of plants and microorganisms which are not ascribed any value. May plants are regarded as weeds and efforts are made to eradicate them from croplands. However, many of the so-called weeds and many microorganisms might be of some distinctive value that we have not so far discovered. Thus, option value refers to the value retaining options available for the future. There could be many plants that could be harnessed for human food in the future and be brought to agricultural practices. There could be some plants we treat and destroy today considering them nuisance weeds, but in the future, they may prove to be novel sources of life-saving drugs. Some microorganisms considered harmful today might be the sources for effective bioremediation of heavy metals and oil spills in the oceans and/or vital for the biodegradation of the so-far non-biodegradable.

Ecosystem Services

Biodiversity is vital to keep natural nutrient cycles going on and sustain food chains, food webs, trophic levels, and energy flows, thus making the ecosystem a self-sustaining unit of nature. The formation, building up, protection, and conservation of soil, water conservation and purification, natural pest control, carbon sequestration, maintenance of gaseous composition of the atmosphere, climate regulation, etc. are the other vital functions nature's biodiversity performs. Further, biodiversity infuses resilience and imparts a high degree of sustainability to an ecosystem and, subsequently, to the entire biosphere.

Ecosystem or environmental services – i.e., the economic notion for ecosystem or environmental functions – the biodiversity performs include (Singh 2009):

1. Supplies of goods (water, foods, fuel, fiber, medicines, etc.),
2. Regulation and control of ecosystem processes, such as the quality of air, water, and climate, erosion, manipulation of biological processes, cause of illness, risk reduction, etc.
3. Cultural aspects, such as non-material benefits that contribute to enriching the quality of life, including cultural diversity, religious or spiritual values, aesthetic values, eco-philosophy, knowledge, inspiration, recreation, eco-tourism, etc.
4. Back-up or supporting activities, such as those required to produce the other services, including primary production, pollination, soil formation, oxygen production, habitat creation, nutrient cycling, etc.

Summary

Biodiversity means the variability among living organisms from all sources, for example, all terrestrial and aquatic ecosystems. It includes varieties of all organisms as well as the ecological complexes they thrive in. Biodiversity prevails in three hierarchal levels: i) genetic diversity, ii) species diversity, and iii) community and ecosystem diversity. The genetic variation within a species (in chromosomal structures, in entire genes, i.e. the traits determining particular characteristics, or in alleles, i.e. different variants of the same genes) is referred to as genetic diversity. The higher the measure of genetic diversity the better the ability of a species to adapt to changed environmental conditions. The evolution of new species – the speciation – depends upon the amount of genetic variability. All the species of plants, animals, and microorganisms, or across all the five living kingdoms in the biosphere count as species diversity.

There are two basic parameters to determine the species diversity of a site or region, viz. i) species richness, and ii) evenness or equitability. Species richness denotes the number of species per unit area. Evenness or equitability denotes the evenness in the number of individuals of a species. The evenness of a species is representative of higher species diversity in a region. The ecological state of a species being unique to a defined geographic location is referred to as endemism. Community diversity reflects variations in the biotic communities in which the species thrive. There are three perspectives relating to community diversity: i) alpha diversity, ii) beta diversity, and iii) gamma diversity. Alpha diversity means diversity within a community. Beta diversity refers to diversity between communities. Gamma diversity means the diversity of the communities or habitats over a geographical area or the entire landscape.

A complex ecosystem provides a large number of ecological niches offering an appropriate range for the activities of the organisms within a community. A dominant species in an ecosystem tends to monopolize natural resources at the expense of others, thus imposing some restrictions on an increment in the biodiversity of an ecosystem. Ecotones or transition areas show a comparatively higher degree of biodiversity, which is an edge effect. Keystone species determine the ability of a large number of other species to persist in the community, thus playing a crucial role in ameliorating the environment for many other species to add to ecosystem diversity. Point diversity refers to diversity at the smallest scale. Epsilon diversity, on the other hand, refers to the diversity of a group of areas of gamma diversities and is also known as regional diversity.

Variations in the natural biodiversity occur following the changes in latitude and altitude. Biodiversity gradually increases from high to low latitude, i.e., from the poles to the equator of the Earth. Biodiversity prevails in high magnitude and abundance in the tropics towards lower latitudes, where ecosystems are more complex, heterogeneous, and stable. It decreases from lower to higher altitudes in the mountains.

Some of the geographic regions of the planet are especially rich in biodiversity. India – a megadiverse country – is one of them. There are 10 biogeographical zones India is categorized into: Trans-Himalayan Zone, Himalayan Zone, Desert Zone, Semi-Arid Zone, Western Ghats Zone, Deccan Peninsula Zone, The Gangetic Plains, North-East India, The Indian Islands, and The Coastal Region.

Biodiversity is of great value to mankind. The direct values of biodiversity include consumptive use value (non-market value of firewood, fodder, medicine, game, etc.), and productive use value (commercial value of timber, fish, crops, etc.). Indirect use values include non-consumptive use value (scientific research, recreation, etc.), option value (value of maintaining options available for the future), and existence value (value of the ethical feeling of the existence of nature, etc.) All foods of all heterotrophs come from living biological resources. Nearly 85 percent of all foods are produced employing cultivated plants. People have been using various plants for the treatment of various ailments. Biodiversity has a great place in our religious and cultural life and has numerous aesthetic rewards for us. Existence value means the value of ethical feelings about the existence of nature. A natural resource as such or a component of it or its biodiversity or ecosystem functioning may not be of any concern or use directly or indirectly. However, the present generation values it, ensuring its availability to future generations. This is called bequest value.

Exercises

A. Multiple Choice Questions

1. Which of the following statements is right?

   a. The higher the measure of genetic diversity the better the ability of a species to adapt to changed environmental conditions.

   b. The lower the measure of genetic diversity the better the ability of a species to adapt to changed environmental conditions.

   c. Both of the above

   d. None of the above.

2. Intra-species diversity is the same as

   a. Species diversity　　b. Genetic diversity

   c. Ecosystem diversity　　d. None of the above

3. Which of the following gives correct information?

   a. Genetic variability within a species increases with environmental variability.

   b. The evolution of new species – the speciation – depends upon the amount of genetic variability.

   c. The higher the number of species in a community, the higher the level of genetic diversity it harbors, and vice versa.

   d. All of the above

4. Species richness denotes

   a. The number of species per unit area

   b. The dry-matter of species per individual

   c. The number and dry-matter of species per square meter

   d. None of the above

5. The ecological state of a species being unique to a defined geographic location is referred to as

   a. Ecological niche　　b. Habitat

   c. Endemism　　d. Competition

6. Diversity within a community is known as
   a. Alpha diversity
   b. Beta diversity
   c. Gamma diversity
   d. Lambda diversity
7. Beta diversity denotes
   a. Diversity within a community
   b. Diversity between communities
   c. Diversity over a vast landscape
   d. Genetic diversity
8. A community is a biotic content of
   a. A species
   b. A population
   c. An ecosystem
   d. None of the above
9. What is the difference between an ecosystem and community diversity?
   a. There is no difference.
   b. The ecosystem comprises both abiotic and biotic components while the community counts only biotic ones.
   c. Community diversity describes a variety of niches, trophic levels, and ecological processes, an ecosystem diversity does not.
   d. A community diversity expresses the features of habitat, an ecosystem diversity of a species population.
10. Which category of the living organisms in an ecosystem is considered a keystone species?
    a. Top carnivores
    b. Primary carnivores
    c. Herbivores
    d. Producers
11. The present generation values it, ensuring its availability for future generations. This value of biodiversity is called
    a. Bequest value
    b. Existence value
    c. Option value
    d. Consumptive value
12. Nearly 85 percent of all foods humanity depends on are produced by
    a. Animals
    b. Fish
    c. Cultivated plants
    d. Dairy animals

13. Which country is especially known for its Marsupials?

a. New Zealand
b. Australia
c. Brazil
d. Uruguay

14. Which of these environments would support the most biodiversity?

a. Amazon rainforests
b. Hydrothermal vents
c. Thar Desert
d. Himalayan Alpines

15. The extreme opposite of endemism is

a. Extinction
b. Cosmopolitan distribution
c. Limited distribution
d. Sparse distribution

16. Alpine pastures are found in the

a. Desert zone
b. Himalayan Zone
c. Semi-Arid Zone
d. Deccan Peninsula Zone

17. What is true about biodiversity?

a. It prevails in high magnitude and abundance in the tropics towards lower latitudes, where ecosystems are more complex, heterogeneous, and stable.
b. Generally, it decreases from lower to higher altitudes in the mountains.
c. It gradually increases from high to low latitude, i.e., from the poles to the equator of the Earth.
d. All of the above

18. An example of the ecosystem value of biodiversity is

a. Nitrogen fixation
b. Soil formation
c. Climate regulation
d. All of the above

19. How many biogeographical zones are there in India?

a. 7
b. 10
c. 14
d. 18

20. Which biogeographical zone of India does occupy the largest area?

a. Trans-Himalayan Zone
b. Himalayan Zone
c. Desert Zone
d. Deccan Peninsula Zone

Answers

1. **a**, 2. **b**, 3. **d**, 4. **a**, 5. **c**, 6. **a**, 7. **b**, 8. **c**, 9. **b**, 10. **a**, 11. **a**, 12. **c**, 13. **b**, 14. **a**, 15. **b**, 16. **b**, 17. **d**, 18. **d**, 19. **b**, 20. **d**

B. Short-Answer Questions

1. How is biodiversity impacted by latitude and altitude?
2. Define genetic diversity.
3. What is species diversity?
4. What do you mean by species richness and evenness?
5. Define Alpha diversity, Beta diversity, and Gamma diversity with appropriate examples.
6. What is the difference between ecosystem and community diversity?
7. Describe in brief and with suitable examples the consumptive use value of biodiversity.
8. What are the non-consumptive values of biodiversity?
9. Define the existence and option values of biodiversity.
10. Name the foods contributed by nature's biodiversity.

C. Long-Answer Questions

1. What is biodiversity? Write an account of all levels of biodiversity.
2. A high level of biodiversity in the tropics is attributable to some of the general characteristics of these geographical areas. Describe with examples.
3. Describe striking features of the various biogeographical zones of India.
4. Describe in detail the benefits of biodiversity to mankind giving suitable examples.
5. Write short notes on the following:

a. Gradients of Biodiversity

b. Biodiversity and Ecosystem Functions

c. Bequest Value of Biodiversity

References

Anubhav, Singh, K., Rastogi, A. and Singh, Vir. 2009. Sacred groves in India: celebrating sanctity of life through biodiversity conservation. In: Gautam, P. L., Singh, Vir and Melkania, U. (eds.) Ecosystem Diversity and Carbon Sequestration: Climate Change Challenges and a Way Out for Ushering in a Sustainable Future. New Delhi: Daya Publishing House. pp. 89-101.

McNeely, J. A., McCarthy, T. M., Smith, A., Olswig- Whittaker, L. and Wikramanayake, E. D. (eds). 2006. Conservation Biology in Asia. Kathmandu, Nepal: Society for Conservation Biology Asia Section and Resources Himalaya.

McNeely, J. A., Miller, K. R., Reid, W. V., Mittermeier, R. A. and Werner, T. B. 1990. Conserving the World's Biological Diversity. IUCN, Gland, Switzerland; WRI, CI, WWF-US, and the World Bank, Washington. 193 pp.

Nautiyal, N., Singh, Vir, Joshi, N. K. and Rastogi, A. (eds). 2012. Climate Change and Biodiversity. Biotech Books, New Delhi. 228pp

Ormsby, A. A. and Bhagwat, S. A. 2010. Sacred forests of India: a strong tradition of community-based natural resource management. Environmental Conservation, 37(3): 320-326. doi:10.1017/S0376892910000561

Shiva, Vandana; Singh, Vir; Dankelman, Irene, Negi, B.; and Singh, S. 2005. Biodiversity, Gender and Technnology in Mountain Agriculture: Glimpses of the Indian Central Himalayas, New Delhi: Navdanya. 90pp.

Singh, Vir. 2009. Environmental services emanating from the Himalayan mountains: valuation against the backdrop of eco-philosophy and chasing the goal of global happiness. In: Gautam, P. L., Singh, Vir and Melkania, U. (eds.) Ecosystem Diversity and Carbon Sequestration: Climate Change Challenges and a Way Out for Ushering in a Sustainable Future. New Delhi: Daya Publishing House. pp. 342-355.

Singh, Vir. 2019. Fertilizing the Universe: A New Chapter of Unfolding Evolution. London: Cambridge Scholars Publishing. 285pp.

Singh, Vir. 2020. Environmental Plant Physiology: Botanical Strategies for a Climate Smart Planet. Boca Raton, USA and London, UK: Taylor and Francis (CRC Press). 230 pp.

Soulé, M. E. and Wilcox, B. A. 1980. Conservation Biology: An Evolutionary-Ecological Perspective. Sunderland, Masschutes: Sinauer Associates.

Wilson, E. O. (ed.). 1988. Biodiversity. Washington, DC: The National Academic Press. https://doi.org./10.17226/989

Wilson, E. O. 1992. The Diversity of Life. Cambridge, Massachusetts: Belknap.

Web Resources

https://en.wikipedia.org/wiki/Biodiversity_hotspot

https://www.easy-grow.co.uk/10-amazing-botanical-gardens-in-the-world/

https://www.environment.gov.au/biodiversity/conservation/hotspots

https://www.gktoday.in/gk/biogeographic-regions-of-india/

https://www.pmfias.com/biodiversity-hotspots-india/

2

Agrobiodiversity Biodiversity in Agriculture

Agrobiodiversity is a key resource for humanity. All of the food retirements humanity depends on are fulfilled by agrobiodiversity. Not only this, fodder, fiber, firewood, and many other basic needs of humankind are also met by agrobiodiversity. Human's dependence on agriculture – not only for survival and sustenance but also for socioeconomic progress, maintenance and celebration of cultural ethos, physical, intellectual, emotional, ethical, and aesthetic development, and psychological contentment and happiness – is inevitable.

Agrobiodiversity is critical for food security for the entire humanity as well as for the food sovereignty of a country. If we are to usher in a sustainable future – that is, a future to be witnessed by healthy, vibrant, and happy humanity (Singh 2019) – there is no choice other than protecting, conserving, and augmenting agrobiodiversity.

Biodiversity, Survival, and Sustenance

All living beings survive, grow, reproduce and sustain due to their continuous nourishment. While all autotrophs – photoautotrophs and chemoautotrophs – are fed by photosynthesis (photoautotrophy) or chemosynthesis (chemoautotrophy or chemolithoautotrophy), all heterotrophs depend on the foods derived from another living organism—autotrophs (herbivory) and/ or heterotrophs (carnivory). Thus, based on the mode of their nutrition, all the living organisms belonging to all the five kingdoms fall into the categories outlined in Table 2.1.

Table 1: Living organisms of all the living kingdoms classified as per their mode of nutrition for their survival and sustenance

Living Kingdoms	Autotrophs		Heterotrophs
	Photosynthetic	Chemosynthetic	
Prokaryotes (Bacteria, Archaea)	√	√	√
Protists	√	-	√
Plants	√	-	√
Fungi	-	-	√
Animals	-	-	√

The human species for its nutrition is all dependent on photoautotrophy. This process is rooted in what we call agriculture. Human beings produce food for their nourishment by managing specific biodiversity in the cultivated area from which they derive nutrients and energy. Animals that human beings use in the socioeconomic systems are also dependent on the biodiversity sustained by photosynthesis.

Agrobiodiversity is not only the exclusive source of nourishment but also the basis of all kinds of livelihoods human beings design and develop. The affluence of agrobiodiversity in our present times is necessary for being maintained and its sustainable conservation and enhancement must be ensured so that its supplies to future generations are ensured.

Biodiversity and Agrobiodiversity

Biodiversity encompasses agrobiodiversity. Agrobiodiversity itself is an integral component of nature's biodiversity. All agrobiodiversity is biodiversity, but all biodiversity is not agrobiodiversity. Agrobiodiversity can be referred to as biodiversity, but the whole biodiversity cannot be referred to as agrobiodiversity. Agrobiodiversity is a notion used for selective biodiversity managed and utilized by human beings for various socioeconomic purposes, especially for foods, fibers, medicine, energy, and numerous other products of specific market values produced through industrial processes.

Biodiversity is counted at three broad levels: ecosystems, species, and genes along with the ecological complexes they prevail in and the ecological processes they prevail through. Agrobiodiversity is in the context of modified ecosystems used for various socioeconomic purposes. These modified ecosystems are croplands, gardens, agroforestry systems, dairy farms, energy plantations, ponds used for producing animal-based foods, aquariums, hydroponics, etc. Agrobiodiversity, thus, encompasses agroecosystems, species, and genetic diversity in cultivated plants, domesticated animals, and microorganisms used in various processes at home and industrial processes. Singh (2005) defines agrobiodiversity as follows: "All land-related activities and processes called agriculture, include a subset of biodiversity that is different from nature's biodiversity, nevertheless very much linked to it and, in fact, an indistinguishable component of it, is called agrobiodiversity."

Nature's biodiversity is laden with very many values but is not generally price-tagged. Agrobiodiversity, however, is price-tagged. In other words, nature's biodiversity is broadly a common property resource and agrobiodiversity a private property resource. Agriculturists, industrialists, planners, and geneticists nowadays are increasingly using the term agrobiodiversity. The term is assuming greater importance because of the attributes agrobiodiversity offers for socioeconomic development. The diversity of agroecosystems is enriched by a variety of farming cultures and agroecological/ agro-climatic zones.

Biodiversity Through the History of Agriculture Development

The history of agriculture had begun with the history of human civilization. Human civilization –that is, a transformation of a distinctive animal species into human species, or humanization of the human species – and changes in the biodiversity patterns in nature have gone hand in hand. Humans' transformation into civilized beings has been a dynamic process: the ancient man had transformed into a modern man over thousands of years. Biodiversity from the wilderness to the cultivated lands has also shifted selectively and dynamically—with the pace of human's exit from forests to modern dwellings. The sort of agriculture we witness today was different at the beginning and in the past (Singh 2014, Singh et al. 2014) and so was the biodiversity we depend upon for our foods.

Have a close look at the human-led development of agriculture from the pre-historic period to date and you will find that it has been gradually deprived of its biodiversity. In other words, agriculture has been recording a gradual reversal of natural evolution – from climax to disclimax.

The Four Eras of Agriculture

The fate of humanity is intertwined with that of agriculture. All land-based human activities encompassing cropping, animal husbandry, horticulture, forestry, fishery, etc. fall under the umbrella term called agriculture. World agriculture has undergone three phases in its entire history and has now ushered in the fourth phase. We can call these phases Agriculture Eras. These are as follows (Singh et al. 2014):

1. Primitive agriculture
2. Traditional agriculture
3. Green Revolution agriculture
4. LPG agriculture

Primitive Agriculture

Primitive agriculture began in the era of the dawn of civilization when man began tilling the land for the first time. The cultivated land was brought into existence through deforestation. However, civilization still depended largely on biodiversity in the wilderness. They derived foods from some 5000 species of plants – herbs, shrubs, and trees. Roots, shoots, stems, leaves, buds, flowers, fruits, seeds, honey, mushrooms, etc. from identified food-yielding plants were included in their foods. A variety of plant-based foods people in the primitive agriculture era incorporated into their diets did not require heat treatment for their processing. However, they had also begun cultivating crops yielding food grains that required cooking. Since primitive agriculture was based on enormous biodiversity, it was a sustainable food production system. This type of agriculture was not linked with the market and represented ideal food sovereignty and a food security system.

Traditional Agriculture

Traditional agriculture had a larger area under cultivated land but had strong linkages with uncultivated (forest/ grassland/ rangeland) areas. As farmers included a larger number of cultivable annuals, largely food grain crops, they depended on a very broad genetic base of every food grain crop. Genetic diversity in traditional farming was a unique example of farmers' art of cultivating agrobiodiversity. Many species of livestock were also reared in the traditional systems. Draught animal power and human muscle power were the only sources of energy (animate energy) for operating agricultural operations. For making the bulk of human foods, the food grains required fuels for food processing. Fuel wood was the exclusive source of fuel for cooking food. Therefore, traditional agriculture inflicted pressure on forests for firewood extraction. This type of agriculture was not strongly linked with the market system. The bartering system was prevailing and it was quite common, especially in rural areas. Because of a broad base of agrobiodiversity, traditional agriculture was also a system with a strong base of sustainability.

Green Revolution Agriculture

Green Revolution, or the Third Agricultural Revolution, emerged following the Neolithic Revolution and the British Agricultural Revolution began in the 1950s and 1960s and was realized by its significant impacts in the late 1960s. The Green Revolution is a set of research technology transfer initiatives that essentially included: i) high-yielding varieties (HYVs) of cereals – especially dwarf wheat and rice varieties with larger grain-straw ratios; ii) chemical fertilizers, iii) pesticides, iv) frequent irrigation, and v) new methods of cultivation, including mechanization and monocultures.

The overall impact of the Green Revolution, a term first used by William S. Goud, the administrator of the United States Agency for International Development (USAID) on 8th March 1968, was considerable shrinkage in agrobiodiversity in the cultivated lands. Increased trend of monocultures led to the decline in species diversity and emphasis on HYVs led to the decline in genetic diversity of selected food grain crops. The exclusion of draught animal power (which was replaced by fossil-fuel-operated machinery) led to the vanishing of many breeds of draught animals.

The Green Revolution type of agriculture has been a great source of environmental pollution. A spurt in agro-chemical based industries necessary to support the Green Revolution also generated the worst sources of environmental pollution. Pollution of soils, waters, air, and even of foods was hardly heard before the advent of the Green Revolution. Pollution of soils owing to new cultivation practices caused depletion in soil biodiversity. This agriculture has strong linkages with the market system. The bartering system has gone down in the history with the vanished agrobiodiversity-rich traditional agriculture.

LPG Agriculture

The liberalization, privatization, and globalization (LPG) era the world ushered in towards the last decade of the 20th century has imparted power to private industry to own and control all natural resources including biodiversity and turn the whole world into a market. Agriculture has also not been spared in the LPG era. The world's agriculture is now in a transition phase—from the Green Revolution phase, now ushering in the LPG era. This agriculture is biotechnology-driven and governed by the global market. This agriculture is largely meant for the profits of the corporate sector. New crop cultivars that nature's evolution 'failed' to add to nature have been created using genetic engineering. Genetically modified organisms (GMOs) belonging to crop cultivars and animals are emerging in research laboratories controlled by big multinational companies. Dozens of Bt crops have been brought into cultivation practices and many more are to occupy cultivated lands in the future. Agricultural scientists often refer to the agriculture initiative as the Gene Revolution, which is a dramatically 'improved' version of the Green Revolution.

LPG agriculture ignores environmental issues and gives no place to agrobiodiversity. Still, in its infancy, this development in the world's agriculture is all set to have unending implications for the environment, ecosystems, and public health. When occupying a significant proportion of cultivated lands, the Bt crops would wipe out the biodiversity of insects, including those playing a crucial role as pollinators. This would eventually culminate in the breaking down of ecological integrity and extinction of several species of pollinators and other insects, many of the crop cultivars, and livestock breeds.

Agriculture is anthropocentric biodiversity management. Nature's biodiversity squeezed out as agrobiodiversity by being brought into cultivation practices began with the beginning of land tillage in the primitive stage of agriculture. In traditional agriculture, agrobiodiversity was in people's hands, i.e., public property. In the Green Revolution era, agrobiodiversity slipped into the private sector resulting in the establishment of several seed companies and many chemical fertilizer and agrochemical manufacturing industries in the private sector. In the modern LPG era, agrobiodiversity is controlled by big biotech companies, and foods are governed by a market system.

The anthropocentric biodiversity/ agrobiodiversity management during different eras has severely affected the ecological state of the resources. While primitive agriculture had its ecological roots and was represented as the most sustainable system of food production in the entire history of agriculture development, the Green Revolution and the Gene Revolution phases of agriculture are quite unsustainable (Fig. 2.1).

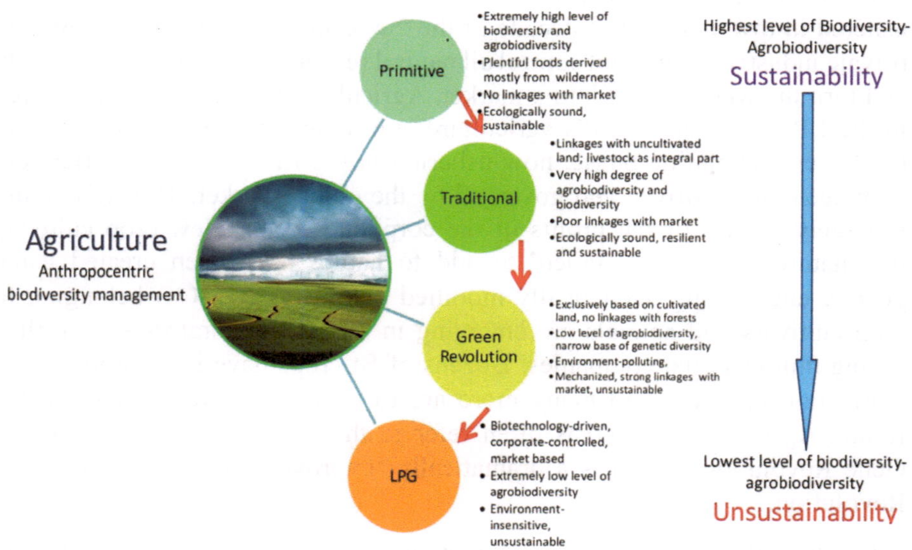

Fig. 2.1: Agriculture through Time's eyes: Anthropocentric biodiversity management from ancient agriculture to LPG agriculture has registered a staggering decrease in agrobiodiversity with a corresponding increase in agricultural vulnerability and unsustainability.

The Industrial Age that began in the 1750s had many things and processes in its "Pandora's box". One of them was the gene-led industrial agriculture, i.e., the LPG agriculture, that came in about 250 years after the beginning of the Industrial Age, and with that, the Industrial Age is all set to assume its climax.

Industrial agriculture, that is, the Green Revolution and the Gene Revolution, is slowly but steadily engulfing the whole system on which agriculture is based, on which humanity imbibes and on which all hopes of life survive (Singh et al. 2014).

Biodiversity and Food Production Scenario

All foods the organisms need are derived from living organisms. All organisms of a species derive their foods from the organisms of certain other species. The greater the degree of biodiversity available for a species the higher the level of food security for that species. An organism's choice of foods is inherently built—determined by its natural evolution. Humans choose a great variety of plants (herbs, shrubs, trees), animals (aquatic as well as terrestrial, invertebrates and vertebrates, insects and reptiles, avian species and mammals), and even microorganisms (bacteria and fungi) for relishing in a great variety of foods. There is no kingdom, no family, no, phylum, and no species from which humans do not choose their foods.

Agrobiodiversity being used in cultivation practices largely varies according to:

i) geographical, or agroclimatic zone;

ii) people's religion, culture, and traditions; and

iii) market values and demands.

Geography, based on its agroclimatic or agroecological specificities, determines the kind of food-producing plants that thrive and can easily be cultivated in a specific region. However, there are several food plants not among the native ones that are being cultivated throughout the world. These introduced plants/ crops in many cases are performing better than even the native food-producing plants and are critical for fulfilling the needs of local people.

People's religion, culture, and traditions are also the major reasons for driving people to cultivate plants and rear animals that have compatibility with their cultural ethos. Some cultures, notably the Hindus, practice vegetarianism, and thus, they cultivate plants and produce foods that fulfill all their nutritional needs. Many tribes in the world depend on forests for their food. Many religious societies choose some specific plants for conservation as per their values. Some cultural groups would grow some specific native plants that provide foods that are especially used on special occasions, festivals, rituals, etc. The Hindus celebrate some sacred notions in their culture, such as sacred forests, groves, sacred trees, sacred plants, sacred seeds, sacred animals, etc. which are pivotal in biodiversity/ agrobiodiversity conservation.

And, of course, market values of certain foods or commercial values of some foods used as raw materials for producing value-added consumable products have also become the prime consideration for cultivating agrobiodiversity in many areas and most societies. This gives way to both native and introduced plants to be cultivated mainly for marketing. In many favorable areas, people give more emphasis on commercial crops rather than on food crops. In our modern world in which lifestyles operate in a cash economy, farmers are compelled to compromise with the native agrobiodiversity and choose the crop cultivars, often non-native, that bring in cash for them.

Biodiversity is the basis of food security as well as food sovereignty. The broader the base of the biodiversity the higher the level of food security and food sovereignty. A huge proportion of biodiversity-rich habitats has been reduced to considerably small habitats which are attributable to indiscriminate deforestation for converting biodiversity-rich lands into cultivated lands. The cultivated lands support only what we call agrobiodiversity, not nature's evolution-led biodiversity. Being reduced to only limited numbers of selected food-yielding and commercial species and the extremely narrow genetic base of each species, the cultivated lands have pushed nature's biodiversity towards extinction.

A new report by the World Wildlife Fund has revealed that agriculture dominated by wheat, rice, and maize is wreaking havoc on wildlife conservation. Not only the wildlife, but these three crops are also threatening the food security of the whole world. Most of the destruction of biodiversity in agriculture has also happened due to these three crops, which are like death knells for nature and food security. According to David Edwards of the World Wildlife Fund, wildlife populations have declined by 60 percent since 1970 due to global agriculture (Singh 2022).

About 75 percent of the food humans consume comes from just 12 plants and five animal sources, with 60 percent of the calories in the entire human diet coming from three crops – wheat, rice, and corn. According to the Food and Agriculture Organization (FAO) of the United Nations, 51 percent of the calories consumed by the world's population come from wheat, rice, and corn alone. Billions of people around the world rely on these crops to survive. Any disruption in their production will have tragic consequences for human civilization.

The World Wildlife Fund in its report titled "Future 50 Foods for Healthy People and a Healthy Planet" (Knorr and WWF 2019) concludes that the conservation of wildlife, and even human survival, is tied to the way we eat the foods. Repeated planting of the same crop in the same area leads to the depletion of nutrients in the soil. Such crops require excessive use of fertilizers and pesticides, which are very harmful to the environment and the consumers of the products of these crops.

Modern methods of agriculture are showing results all over the world. There has been a staggering decline in many species of wildlife in the last 50 years. Insect species have caused great destruction in many areas, due to which pollination of plants has started disrupting. Modern agricultural systems have pushed wildlife to an end. If the insect species continue to disappear at this rate, millions of plants will become extinct from the earth and agricultural production, and consequently food security, will be in jeopardy.

In ancient times, we used to choose our food from thousands of plant species, but in the era of the Green Revolution our food has been reduced to some three dozen plants, and among these also the breath of all mankind is stuck only on wheat, rice, and maize. What an irony that human civilization is moving towards food poverty with time! There used to be thousands of varieties of these three crops in traditional farming. For example, there were some 60,000 varieties of rice in our country itself, but the Green Revolution, serving only the purpose of high productivity, also devoured the genetic diversity of these three crops. Now only a few cultivars with "maximum yield" are cultivated and thousands of varieties have become extinct and some are 'preserved' in the frozen temperatures of gene banks.

If the three crops taking care of humanity's wellbeing were destroyed due to famine, plant disease, pest infestation, climate change, or any other mishap, then what would happen to mankind? Anyway, by the year 2050, to satisfy the hunger of one-third of the extra stomachs in the world, 70 percent additional food production will be needed, which these three crops will not be able to supply.

The European Union, China, and India are among the largest wheat producers accounting for 20 percent, 18 percent, and 13 percent of the world's total wheat production, respectively. China, India, Bangladesh, Indonesia, and Vietnam top the world in rice production, accounting for about 24 percent, 19 percent, 7 percent, 6 percent, and 5 percent of the world's total annual rice production, respectively. The US, China, and Brazil account for 32 percent, 22 percent, and 10 percent of the world's total maize production, respectively. America is the largest exporter of corn.

According to the Food and Agriculture Organization (FAO) of the United Nations, about 75 percent of biodiversity has been lost in the agricultural sector. There are over 2000 varieties of apples in the UK, but no more than 20 will be found in supermarkets. In our contemporary world, agriculture is no longer aimed at ensuring food security directly. It is controlled by the demands of the market. What the market wants, not the people, grow in the fields. The key to food security is also in the hands of the market. Destroying the diversity of food-giving plants and giving place to only a few plants in Green Revolution agriculture and leaving the fate of human civilization in the hands of only three crops – is this not a negative contribution of Green Revolution to humanity?!

To adapt our agriculture according to the principles of environmental protection and to transform the food according to the socio-cultural objectives of physical, intellectual, moral, emotional, and psychological development is the greatest need of the present to secure the future of human civilization. The discovery of nutrient-rich plants has prompted us to return to ancient cereals and heritage plant varieties. It is best to try to find some forgotten plants again and add them to the food. Avoiding food dependence on meat, poultry and fish will improve public health as well as control climate change.

There has been homogeneity in the diet of all humans around the world, which is the root cause of the shrinking biodiversity in the agricultural sector. There will be no human who does not depend largely on any of the three types of grasses—wheat, rice, and maize! South Asia runs on the energy of rice. "Dal-Bhat" is a major indicator of happiness in Himalayan states in India and Nepal.

The energy with which we think, discharge all our actions, earn fame, can leap to the Moon and Mars, and the energy which gives us the ability to explore all the galaxies is obtained from seeds, and the seeds of the three species we get that energy from belong to the Poaceae family. The essence is that our civilization survives only based on the three plant species of the Poaceae family.

Pyramid of Agriculture

A current picture that emerges from the aforesaid discussion, especially from the background from which Fig. 2.1, clearly reveals that the pyramid of food production is straight or upright suggesting that there is enough food in the world that is produced following modern agricultural practices; however, the conceived ecological pyramid is an inverted one (Fig. 2.2). The inverted ecological pyramid indicates that the food production capacity of the agroecosystems is almost squeezed by intensive cultivation management and induced ecological vulnerability. The upright food pyramid indicates the state of sustainability, but this state is only indicative of the state just before the ecological collapse.

(a) (b)

Fig. 2.2: Food pyramid **(a)** and Ecological pyramid **(b)**: The straight food pyramid indicates plentiful food supplies but at the cost of an ecological disaster reflected in the inverted ecological pyramid. A sharp contrast between the two pyramids is indicative of an unsustainable food production system.

We have discussed the four phases of agriculture development, which represent four different eras of the state of agriculture (see Fig. 2.1). If these eras are represented through a pyramid, a picture like that in Fig 2.2 would emerge. The base of the pyramid occupies primitive agriculture with maximum photosynthetic efficiency, maximum biomass, and varieties of plentiful food. Most of the solar energy in living form (in the green plants) dwells in this agriculture flourishing with natural forests, limited cultivated crops, and livestock. There is enormous biodiversity (or agrobiodiversity usable for foods and feeds) in this type of agriculture that blossoms amidst ecological affluence. Ecological affluence is created by solar energy transformed into living energy – the energy of life. Solar energy is transformed into living energy (biochemical energy) through photosynthesis. Photosynthetic efficiency would be at its highest level when there is as much biodiversity in a system as nature has evolved over millennia. Since primitive agriculture flowers on the natural biodiversity of nature, it naturally occupies the base of the agriculture pyramid, making the whole life strong, vibrant, extremely resilient, and sustainable.

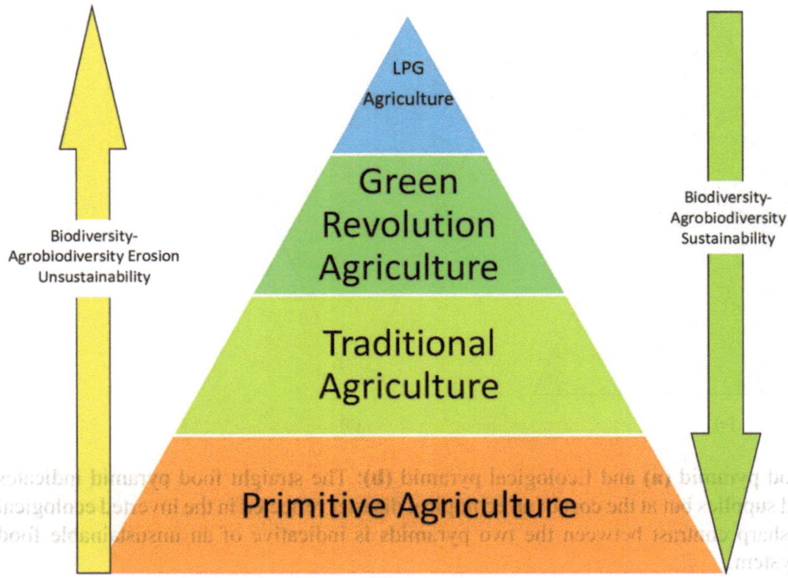

Fig. 2.3: The Pyramid of Agriculture: Energy, biomass, resilience, and sustainability decrease towards the tip of the pyramid, and vice versa

The next broader base upward the agriculture pyramid is occupied by traditional agriculture. Primitive agriculture depended mostly on innumerable varieties of uncultivated foods occurring in natural forests. On the other hand, traditional agriculture depends upon innumerable varieties of cultivated food crops. Linkages of croplands with natural forests/ rangelands/ grasslands and dependence on uncultivated foods also exist. Livestock are used in the transfer of nutrients from uncultivated areas to cultivated ones and in managing nutrient recycling. They are also used in diversifying agriculture, generating draught power, and in producing fiber and foods (especially milk). Livestock are an integral part of this agriculture and contribute to enhancing ecological integrity. Owing to extreme agrobiodiversity managed in the cultivated areas and biodiversity in uncultivated areas, this agriculture is also quite energy-efficient, but less so in comparison to primitive agriculture. As traditional agriculture works closer to the base of the pyramid, it makes life ecologically strong, vibrant, resilient, and sustainable.

The Green Revolution agriculture is cursed to rely upon monocultures, which are extremely vulnerable and unsustainable. From the energy efficiency point of view, this agriculture is less efficient, for it ignores agrobiodiversity and has no relationships with forests and other uncultivated areas. Ecologically disintegrated, vulnerable, polluting, unhealthy, and unsustainable, the Green

Revolution is a compromise with the very sustainability of life. Therefore, not much living energy resides in the Green Revolution agriculture, and whatever energy is there, it cannot support life on a sustained basis. It cannot carry humanity into a future full of hopes. Therefore, it lies towards the tip of the agriculture pyramid. Further, it poisons the environment with the use of deadly pesticides and chemical fertilizers, to the overall unsustainability of life.

The LPG agriculture spells ecological disintegration to be followed by disastrous consequences. There is a complete wipeout of genetic diversity with a limited number of genetically modified or transgenic crops (crops with alien genes, such as poisoning bacteria) replacing the green revolution crops. Transgenic crops to realize their full production potential need all sorts of chemical fertilizers, but avoid the use of pesticides, for the synthesis of poison due to the presence of an alien gene, which takes place inside a plant. All sorts of insects prospering on vegetation would be destroyed if they harbor a transgenic crop. The transgenic crops are potent enough to wipe out all varieties of insects, including the pollinators so vital for the reproduction of most of the vegetation, triggering the process of ecological disintegration. Occupying the tip of the agriculture pyramid, this agriculture costs vibrancy, resilience, and sustainability of whole life.

Agriculture and A Sustainable Future

In agriculture, there is no sudden shift from one system to the other. A system evolves with time. Farmers have been changing their agricultural systems from time to time and according to specific circumstances. There has been a good deal of diversity among agricultural systems. Different systems evolved in different geographical settings by different cultures. Primitive agriculture and traditional agriculture encompassed enormous diversity of farming systems. Therefore, agriculture under these two broad categories is extremely resilient and sustainable and ensures us a sustainable future.

The future of the time is inevitable. It is imminent, whether with humanity or without humanity. But a sustainable future should be witnessed by happy and blossoming humanity. There are two preconditions for happy and blossoming humanity, that is, of a sustainable future: (1) availability of plentiful food; and (2) ecological affluence. Availability of plentiful foods implies a healthy, vibrant, environmentally safe, ecologically ameliorating, resilient, and sustainable agriculture. Ecological affluence implies natural resources in a healthy and regenerative state, biodiversity at its climax, and human beings in a healthy relationship with all other species of life.

The availability of plentiful foods and ecological affluence are inseparable. Ecological affluence ensures plentiful foods obtainable from innumerable plants, cultivated as well as uncultivated. Ecological affluence, on the other hand, is ensured by pursuing, following, and sustaining the most appropriate food production systems. And a sustainable future is ensured by a healthy and ameliorating relationship between the two.

The two categories of agriculture, viz., Green Revolution and LPG, cannot carry us to a sustainable future, for they are based on shrunk agrobiodiversity and are polluting, degenerative, and disastrous. On the other hand, the other two categories of agriculture, viz., primitive agriculture and traditional agriculture, are life-enhancing and, therefore, ensure humanity ushers in a happy and sustainable future.

Agriculture Pyramid and Fate of Humanity

There are certain geographical areas and specific cultures still operating in primitive and traditional agriculture. Mountains of the Himalayas, tribals in almost all Indian states and many areas, and tribes in Africa and Asia are still maintaining their identity with primitive and traditional agricultural systems. There are still numerous tribes who do not plow their fields for growing foods but pick up foods from edible mushrooms, roots, shoots, leaves, buds, flowers, pods, fruits, honey, etc. found plentifully in the natural forests they inhabit.

There are many areas in the world where farmers are still managing traditional practices in agriculture by cultivating land using draught animal power. They continuously enrich their soils with organic fertilizers and cultivate a large number of landraces of all food crops – millets, cereals, pulses, oilseeds, fruits, vegetables, condiments, spices, etc. They use traditional methods of plant protection against insects and disease pests. They do not use external inputs in their fields. Their approach to food production is farming system based and they maintain an appropriate ratio between uncultivated land (forests/ rangelands/ grazing lands) and cultivated land and rear many species of livestock – cattle, buffalo, sheep, goats, etc. – as part of their farming systems and farming cultures.

The above-mentioned farming cultures in some inaccessibility-ridden areas have so far kept themselves impervious to any significant change in their agriculture. However, institutional intervention involving government departments, research institutes, agricultural universities, etc. is too intensive that their agriculture is not purely primitive or traditional. Many changes have occurred so far; for example, forest-dwelling cultures have begun plowing land, and traditional farmers have been compelled to cultivate so-called high-yielding varieties (HYVs) developed by research and corporate organizations. The HYVs are HRVs – the high-responsive varieties. The HRVs cannot perform without adequate chemical fertilizers, pesticides, and irrigation. With a switch over from landraces or traditional varieties of seeds to HRVs, traditional farmers have been compelled to depend on the market for buying costly chemical fertilizers and pesticides of all sorts. And now these farmers in many areas are being forced to cultivate transgenic crops which could spell out ecological disaster.

Again reminding our agriculture pyramid (Fig. 2.3), we can predict how the four major categories of agriculture we have referred to as the agriculture era are going to support human populations in the future. Foods nourish us with nutrients and energy. Applying ecological principles, as we have seen in Fig. 2.1, most of the nutrients and energy are trapped in the vegetation – the producers, or the first trophic level – occupying the base of the ecological pyramid we have discussed. Since most of the nutrients and solar energy-transformed-into-biochemical energy lie in this trophic level, it would be capable of supporting the largest populations. In other words, as per the principle of ecology, the most stable populations are those living closer to producers—the photosynthesizing biodiversity. Since the largest proportion of biomass is encompassed in the producers at the base of the ecological pyramid, the primitive sort of agriculture would support the largest human population. Thus, the future of humanity is safest by depending on this agriculture.

In the ecological pyramid, the next trophic level is constituted by herbivores. Traditional agriculture can be equated with herbivores, for this agriculture consumes natural vegetation by transforming forests and other uncultivated areas into cultivable fields. The next largest store of solar energy and nutrients is in the herbivores. Thus dependence on traditional agriculture will ensure the well-being of a pretty large human population. Thus, a future dependent on this agriculture would be quite safe, but not as safe as depended on the primitive one.

The Green Revolution agriculture *behaves* like carnivores, the next trophic level in an ecological pyramid. It consumes much of the biodiversity of traditional agriculture, thus limiting the potential of photosynthesis the foods-producing plants perform. Thus energy and nutrient level is reduced considerably at the level of the Green Revolution and, therefore, it can support only a limited population or a large population for a limited period.

LPG agriculture behaves like top carnivores. This agriculture has devoured (or will do so) most of agrobiodiversity replacing it with fewer transgenic crops. This trophic level lying towards the tip of the ecological pyramid can be equated with the top carnivores, like a lion, tiger, cheetah, vultures, etc. who are not left with the choice of varieties of plentiful foods to consume. Thus, many top carnivores are on the verge of extinction. What would be the fate of humanity should it depend on transgenic foods of the LPG agriculture, can, thus, easily be predicted.

Mountain Agroecosystem Specificities

Mountain areas are altogether distinguishable from those of the plains. Distinctive characteristics or specificities emanate from the mountain perspective framework evolved by Jodha (1990). Mountains share six common characteristics, viz., marginality, limited accessibility, fragility, diversity or heterogeneity, ecological niches or comparative advantage, and human adaptation mechanisms (Fig. 2.4). The first three characteristics are the negative attributes of the mountains, whereas the last three are the positive attributes.

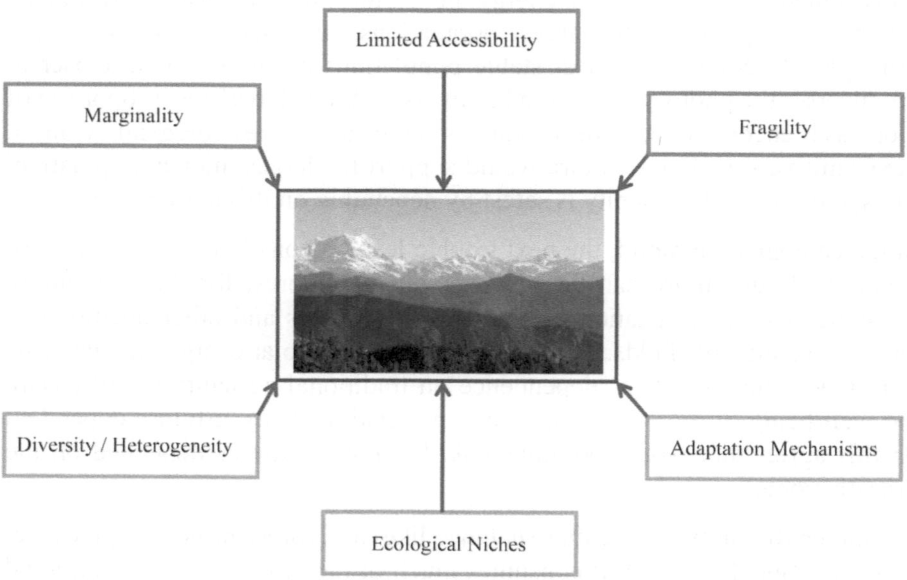

Fig. 2.4: Mountain ecosystem characteristics distinguishing mountain agroecosystems from those of the mainstream plains. The three specificities on the upper side of the Figure reflect negative attributes and the three on the lower side offer positive attributes derivable through development intervention.

The above-mentioned mountain ecosystem specificities determine the level of biodiversity in the region. A brief description of the mountain specificities is given below (after Jodha 1990, 1993, Jodha et al. 1992):

- **Marginality:** It is one of the most known mountain specificities which means a characteristic that counts the least concerning 'mainstream' situation. Basic factors contributing to mountains' marginality are remoteness/ physical isolation, fragile or low-productivity resources, and several man-made handicaps.
- **Limited Accessibility:** This is attributable to slope, altitude, overall terrain conditions, and periodical seasonal hazards (snowfall, landslides, avalanches, etc.).
- **Fragility:** It is attributable to altitude and steep slopes, in association with geologic, edaphic, and biotic factors, which limit the mountains' capacity to withstand even a small degree of disturbance. When mountain resources and the environment begin to deteriorate due to any disturbance, the effects of fragility occur at a faster rate.
- **Diversity or Heterogeneity:** Extreme degree of diversity or heterogeneity in the mountains is a function of interactions of different factors, such as elevation, altitude, geologic and edaphic conditions, steepness and orientation of slopes, wind and precipitation, mountain mass, and relief of terrain.
- **Ecological Niches:** These reveal a situation of comparative advantage in the mountains which is owing to their specific environmental and resource-related features due to which mountains provide a 'niche' for specific activities or products unique to mountain ecosystems. At the operational level, mountains may have a comparative advantage over the plains in these activities.
- **Adaptation Mechanisms:** This specificity has been evolved by mountain communities through trial and error over generations. Adaptation mechanisms or experiences are reflected through formal and informal arrangements for the management of resources and diversified and interlinked activities.

Mainstream Agriculture Vs Mountain Agriculture

The Himalayas comprise the youngest, the highest, and the most fragile geological formation on Earth. Fragile mountain ecosystems are found embracing a very high degree of biodiversity—alpha, beta, and gamma diversity. Mountain ecosystems in the Himalayas are encountered playing a very interesting role largely unmatchable with the mainstream plain areas. The Himalayan ecosystems are distinctive from those of the plains in many of their characteristics. Global agriculture has undergone massive changes, including those in agrobiodiversity patterns. However, farmers in the Himalayan mountains have not transformed their agriculture to a significant extent. Looking at the high degree of inaccessibility and marginality and fragility features, there seems little scope for agricultural transformations in the mountains (Singh 2018). Mountain agriculture still maintains its traditional features to a great extent. Cropping systems tend to cultivate agrobiodiversity following farming situations (for example, altitude, slope aspect, irrigation facilities, etc.). Some of the key features of mainstream and mountain agriculture are shown in Table 2.1.

Earlier, Jodha (1990), Jodha et al. (1992), Singh and Tulachan (2002), Singh et al. (2014), and Singh (2018) have attempted to portray distinguishable features of mountain agriculture. Comparative analysis of the two contrasting agricultural systems is based on these mountain scholars.

Table 2.1: Some contrasting features of mainstream and mountain agriculture

Features	Mainstream agriculture	Mountain agriculture
Farming system	Absent	Present
Fragility	Moderate	High
Marginality	Low	High
Inaccessibility	Least	High
Diversity/ heterogeneity	Minimum	High
Niche/ comparative advantage	Poor	Rich
Adaptation capability	Poor	High
Complexity	Less	High to extreme
Vulnerability	high	Moderate to high
Resilience	Low	High
Water use efficiency of crops	Poor	High
Inputs	External	Internal
Productivity	Moderate to high	Moderate
Sustainability	Poor	High

Agroecosystem Diversity

Agriculture is often isolated as cropping activity taking place on cropland cut off from other components of nature. Agriculture includes all land-related activities, such as cropping, forestry, horticulture, seed production, animal husbandry, etc. An agroecosystem involves croplands, forests, grasslands, agroforestry systems, livestock, water sources, etc. Those components of nature, such as a forest or grassland or a water source that are not used for farming culture directly or indirectly are not a part of an agroecosystem.

An agroecosystem is an arbitrarily defined unit of nature with distinct energy flows and chemical cycling, involving uncultivated land areas (forests, grasslands/ grazing lands), cultivated land, and livestock in organic linkages with each other, woven into a complex unitary whole, functionally oriented to produce foods and other life-supporting products – such as feed, fiber, fuel, fertilizers, etc. – while maintaining ecological equilibrium. An agroecosystem overlaps diverse ecosystems designed and managed by a farming community for the production of fundamental life-sustaining products.

Uncultivated land area – forests, grasslands, grazing lands – is a core component and the whole biodiversity conserved in this component nourishes the cultivated land through the constant flow of nutrients. Cultivated land is managed through ethno-engineering often by clearing original forests. Livestock are the dynamic component of an agroecosystem; nutrient flows from the uncultivated land are mediated by livestock. A community of farmers living in a cluster of houses called a village manages an agroecosystem.

Uncultivated land flourishing with the diversity of vegetation imparts ecological stability to the whole food production system. A forest also enhances resilience, complexity, progressive succession, ecosystem efficiency, productivity, and eventually sustainability of the farming system.

The agroecosystem should be regarded as a foundation for conceptual and practical approaches to the sustainable development of agriculture. The uncultivated land–cropland ratio is crucial for the ecological sustainability of an agroecosystem. It should be decided as per the agroecological zone in which an agroecosystem lies. An agroecosystem should not comprise more than two-thirds of the cultivated area in the plains and one-third in the mountains.

Farming operating in such an agroecosystem will be ecologically sound and least vulnerable to damage by abiotic and biotic factors. Biodiversity in the uncultivated and cultivated components of an agroecosystem would be saved, conserved, and enhanced. There would be a plentiful supply of nutrients from the uncultivated land to the cropland and the farming system would by and large restore its self-containment features. Moisture circulation in the atmosphere and water conservation within the soils of an agroecosystem would be ensured.

A biodiversity-rich mountain agroecosystem constantly enriches the cropland soils through internal nutrient flows and recycling in the cultivated lands. In such a system, the soil supports a healthy population of earthworms, bacteria, fungi, actinomycetes, protozoa, millipedes, insects, and a host of other creatures that are essential for sustaining the health of the soil and improving the mobility of the soil nutrients from the soil to the plant root system, a condition that helps the system enhance and sustain high productivity.

In an agroecosystem approach to food production, the soil is not regarded as an inanimate substratum but as a wonderful ecosystem. The productivity of a system is a function of the soil ecosystem. A soil ecosystem reverberating with a bounty of life forms (microflora, microfauna, mesofauna, macrofauna) will have bearing on the health of life forms in the troposphere, including that of plants and human beings.

Much more food would be produced from such a system than it would have been produced from merely the cropland, for a variety of foods would be available from the uncultivated area also. The system would be of great economic significance, for the farmers would earn not only from cropland but also from the produce of uncultivated lands.

Demand on the market for the purchase of inputs would be reduced and, in the long term, eliminated. There would be no agriculture-borne pollution of the environment. Sound health of the system, livestock, and people would be ensured. The system would promote social-cultural cohesion amidst the community, for the uncultivated lands can then be used as common property resources (CPRs). The ecological integrity of the farming system would be maintained and the sustainability of the agriculture would be ascertained.

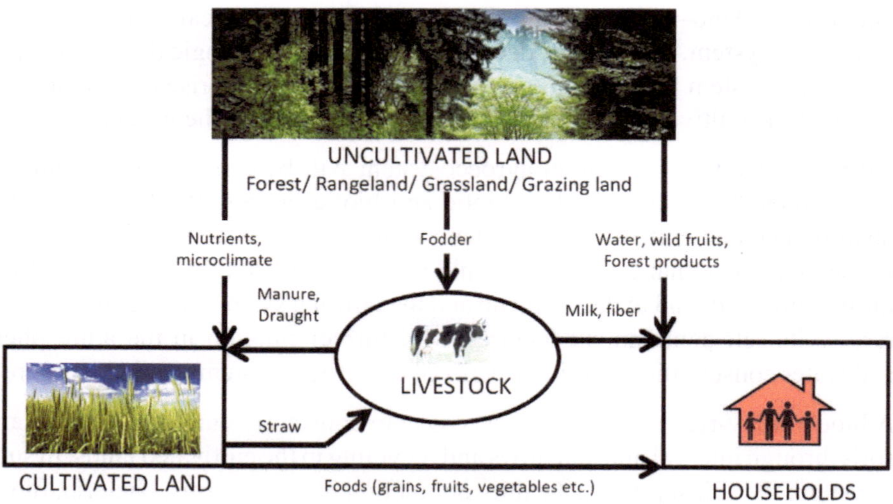

Fig. 2.5: An integrated mountain agroecosystem showing matter and energy flows between components

A typical agroecosystem with uncultivated and cultivated lands, livestock, and households linked with each other through nutrient flows is shown in Fig. 2.5. The uncultivated land which is often under the reserve forests, community forests, rangelands, grazing lands, and such categories of land, plays a key role in imparting essential ecological stability and integrity to the whole system and also ensures plentiful supplies of a variety of foods from uncultivated and cultivated lands, livestock resources and, wherever possible, from water resources. Such an agroecosystem is self-contained, i.e., no input needs to be imported. However, the system is capable to export food and render vital environmental services (e.g., water flows, oxygen, optimum weather, etc.) even to far-flung plain areas.

Uncultivated Land

Most of the land in the mountains is uncultivated land. Various types of uncultivated land are forests, rangelands, grazing land, grassland, marginal land, abandoned cultivated land, open spaces, homesteads, etc. In fragile mountains, the ecological stability of the system should be the foremost criterion for agricultural sustainability (Singh 2005). The natural forests harboring a great measure of biodiversity make an agroecosystem more vibrant and functionally more efficient. Forests also serve as the basis of the sustainability of the whole agroecosystem.

A forest's contribution may be both tangible and intangible, in terms of products and components of an agroecosystem. In the form of directly extractable products, a forest provides fuel wood, timber, fodder, fruits, food, etc. and in terms of a component of an agroecosystem, a forest's contributions are in the form of biodiversity, moisture circulation, nutrient cycling, and the make-up and stability of environmental and physical resource variables. These functions of a forest are the system's inputs and contributions to the sustainability of an agroecosystem.

Fuel wood is the most important source of energy in rural mountain areas throughout the length and width of the Himalayas. Many families depend on fuel wood for cooking their food. Alternative energy sources are not easily accessible or are too expensive. The use of wood as a source of cooking energy also contributes to managing soil fertility indirectly, for in the absence or paucity of it the animal dung, instead of being used as manure, would be dried to be used for cooking food at the expense of soil health.

Conventional agriculture, dominating the plains, demands constant supplies of energy, e.g., fossil fuel for operating machines and transportation of the produce. In the agroecosystems in the mountains, nevertheless – owing to almost no mechanization – the petroleum fuel is hardly in demand. The need for soil nutrients is met instead by farmyard manure and through nutrient recycling management mediated by livestock in the mountain farming systems.

In the fragile mountains, ecological amelioration of the farming system should be the foremost criterion for sustainability. A forest supplies a massive amount of nutrients to livestock and cropland. It also plays a phenomenal role in the hydrological cycle. Owing to its contribution to the processes and functions leading to the ecological sustainability of the mountain farming systems, a forest continues to be an essential feature of mountain agriculture, though much degradation of forests has already occurred in many areas, and such areas are facing grim consequences.

In farming systems, deep-rooted woody perennials contribute to soil organic matter through the shedding of their leaves. This organic matter improves soil structure, fertility, and water holding capacity as well as ameliorates the functioning of the soil ecosystem. Deep-rooted trees absorb nutrients from greater depths and bring them onto the topsoil. Thus shallow-rooted crops are nourished by these nutrients met by deep-rooted ones.

Contributing as windbreakers, the trees also increase moisture availability by reducing wind and, thereby, lowering evapotranspiration rates. Tree canopies can lower the impact of heavy rainfall, reducing runoff and increasing water filtration into the soil. Moreover, the shade provided by trees lowers soil temperatures reducing evaporation and slowing down the decomposition rate of organic matter. Leguminous perennials may provide still further advantages by adding nitrogen to the soil.

Trees, as contributors to biodiversity and natural flows, facilitators of regenerative processes, and producers of diverse biomass, can play multiple roles in resource intensification-centered approaches to sustainable land use (Jodha et al. 1992). A diverse and complex forest ecosystem, apart from enhancing the productivity level, imparts a great degree of resilience to a food production system and ensures farmers' security.

In addition to maintaining the quality of the resource base, the trees contribute to the vitality of the entire agroecosystem, which is best ensured when the soil is well-managed. The health of plants, animals, and people is maintained through biological processes (Reijntjes et al. 1992). Since renewable plant diversity and recycled nutrients in the self-contained mountain farming systems are the basis of soil fertility management, and natural biological protection of plants, no pollution in the system is caused in that the use of eco-malignant pesticides is avoided. Thus, a forest, or assemblage of trees, provides a base for ecologically sound and environmentally safe agriculture.

Forest management practices involve the use of an ecosystem's produce as a subsidy for the other component (e.g., forest biomass is used as livestock feed and manure for cropland). A forest thus helps strengthen linkages amongst various components of the farming system. This contribution is vital for maintaining the ecological integrity of the whole system. Forest's role in biodiversity, natural regeneration, nutrient flow and recycling, moisture circulation, building and binding of soils, etc. is also vital for infusing sustainability into the whole system.

Cultivated Land

The uncultivated–cultivated land ratio in the Himalayan mountains is very large. Only about 10 percent of the land is cultivated in Uttarakhand, for example. Minimum terracing of mountain slopes for raising food crops is important for managing agroecosystems in fragile ecosystems. In some areas, intensive terracing is also evident which is not conducive to specific mountain conditions.

Terraces on gentle slopes by traditional mountain farmers are a wonderful form of ethno-engineering. Almost all terracing work has been carried out manually with little aid of draught and carry-pack animals. All operations of crop cultivation are carried out by using draught animal power particularly provided by bullocks, the male cattle. These operations are land preparation (plowing and leveling), puddling, inter-culture operations, threshing, etc. Human labor, particularly women's labor, is also expended to a great extent.

There are two broad categories of cultivated land: Irrigated and unirrigated (or rainfed). In the valleys drained by rivers, streams, and small rivulets, the cultivated land is almost plain. In the uplands – mid- or lesser Himalayas, high altitudes, and greater Himalayas – the cultivated land is largely rainfed. The rain-fed terraced fields are used for raising crops without irrigation, and the cultivation practices in these areas essentially involve agroforestry.

Cultivated land in different microclimatic zones raises different crops, different varieties of crops (often landraces), different cropping systems, and different cropping patterns. Cultivated land-uncultivated land linkages are maintained naturally. For example, nutrients from forests to cultivated land flow through winds and rains and by humans and livestock.

Livestock

Livestock acquire special importance in mountain farming systems both on ecological and socio-economic grounds. They are an integral part of the farming system and a lively 'bridge' connecting two types of land, viz., the uncultivated forest land and cultivated land. This linkage is crucial for the ecological and socioeconomic sustainability of the system. Forests, especially the natural ones, are a rich repository of nutrients, which become a subsidy for cultivated land. These nutrients are transferred to the cultivated land via the agency of livestock. Nutrient transfer takes place in two ways. Forest biomass – tree leaves and ground flora – are fed to livestock. Biomass is also used as bedding material in livestock sheds. Both dung and bedding material are converted into manure which is transferred to the cultivated land. Livestock also recycle the nutrients in the cultivated land. Crop residues are fed to the animals and thus the nutrients in them are recycled into the cropland.

The biomass transfer and cyclic flow pattern of nutrients mediated by livestock infuse vitality in the production system and livestock themselves fulfill their requirements for maintenance and production (milk, draught power, etc.). This relationship between a forest ecosystem, livestock, and cultivated land, as we have already mentioned, is vital for the sustainability of mountain agriculture.

As indicated above, livestock are interposed in this flow pattern to capture some of the energy and transform it into work and milk. Animals are not essential to the system, for vegetation from uncultivated land and crop residues can be composted directly, as they are to an extent in Chinese agriculture. Composting, however, is wasteful of energy. Livestock recover some of the energy otherwise entirely lost in the composting process (Singh and Jackson 1992).

Two common livestock raising systems are the sedentary and transhumance system. Animals kept under a sedentary system are kept in and around a village throughout the year. All animals, except buffaloes, are kept on daytime grazing. At night, crop residues and tree leaves are provided to them. Buffaloes and crossbred cows are mostly stall-fed. Buffaloes and cattle are the preferred species. Some families keep few to a large number of goats.

Under the transhumance system, animals move to different locations which mainly depend on crop seasons. Sheep and goats are the most preferred species managed under this system. All animals during summer when they are herded on high altitudes, mainly in alpine meadows, virtually stay on grazing. During the winter months, animals move to the lower altitudes. The nomadic herders generally do not have permanent houses or cultivate crops, while some tribes practicing transhumance, have their cultivated land and also practice cropping under the traditional livestock-based farming system.

Buffaloes are the preferred milch animals in the mountains. Cows are mainly kept to produce bullocks. Buffaloes are also given preference in the feeding of high-quality feed items, particularly in the sedentary system of animal keeping. Bullocks are provided special feeding care during the plowing season and also on special occasions when the importance of *goudhan* (the cattle wealth) is celebrated.

Livestock size and composition are dictated by the ecological and socioeconomic realities of the region. Cattle size is determined mainly by draught requirements. People's reverential attitude (or sentiments) towards cattle is not a factor affecting livestock population composition as sometimes misconceived. It becomes clear from the fact that over the last many decades buffaloes' and goats' numbers have been increasing while those of cattle are constant or decreasing (Singh and Tulachan 2002). Cattle husbandry in the Central Himalayas revolves around the requirement of agricultural draft and dung for manure and the small yet sturdy and sure-footed local animal breeds serve the purpose. The agro-pastoral society developed a lasting farming system with a high degree of husbandry in animal breeding, crop production, and land management.

Most of the cattle in the Himalayas are small and nondescript. The crossbred population of cattle accounts for five percent of the total cattle population. Buffalo population is largely composed of the grades of Murrah from the neighboring plain states. Sheep are mostly concentrated in the upper reaches of the Himalayas. Goats are widely distributed in all districts. Cattle bullocks are the sole source of draught power in mountain areas almost everywhere.

As manurers, exploiters of wastes, sources of power, risk cushions, transporters, and sources of raw materials (milk, meat, wool, etc.), the livestock benefit the mountain farmers directly. Furthermore, livestock play an important role in the cultural identity of mountain people.

Households

Households generally are not regarded as part of an agroecosystem. But they are an integral part of it. Agriculture is a play of energy and nutrient capture. Where do most of the nutrients go to? They go to households to be consumed by household members. Humans are the major consumers of the energy and nutrients trapped by crops. The energy and nutrients that come from soil to households are critical not only for a family's welfare but also for the productive performance of agriculture.

Part of the energy transferred to households through food is expended back into agriculture through human labor. Energy transferred through feeds (straws, fodder, weeds, and food grains) is expended into agriculture by employing draught animal power.

In households, nutrients are there in kitchen waste, human waste, and animal waste. These nutrients must be recycled into the soil from where they have come. Management of the nutrient flow between cultivated areas and households is a matter of deep consideration for the sustainability of food production. When this flow is not managed with effective measures, nutrients into cultivated land have to be supplied from outside, i.e., have to be bought from market. Dependence on external inputs creates many side-effects and after-effects in agriculture. Therefore, a household serves as an integral component of an agroecosystem in the mountains. Further, a household itself comprises its cropping system; we can call it the household-based cropping system. This system cultivates vegetables and fruits for home consumption. Occasionally ornamental plants are also cultivated as part of the household-based cropping system. It is different from backyard farming or kitchen gardening, for it is a part of the whole farming system or agroecosystem. In the plains where not-so-common kitchen garden or backyard farming is practiced, inputs are bought from the market. In the mountain only internally produced nutrients are the basis of agriculture.

Wild Relatives of Crop Plants in the Himalayan Region

The Himalayan Mountain Region has provided an appropriate environment for several varieties of cereals, pulses, fruits, oil-yielding plants, spices, tuberous vegetables, and sugar-yielding plants and their close relatives that number 155 (Arora and Nayar 1984). A list of wild relatives of economic plants in the Himalayan Region is given in Table 2.2.

Table 2.2: Wild relatives of economic plants in the Himalayan Region

A. North-West and Northern Himalayas	
Cereals and Millets	*Aegilops tauschii, Avena barbeata, A. fatua var. fatua, A. ludoviciana, Digitaria sanguinalis, Elymus dalutricus, E. dasystachys, E. nutans, E. distans, E. orientale, Hordeum glaucum, H. spontaneum, H. turkestanicum, Penisetum orientale.*
Legumes	*Cicer microphyllum, Lathyrus aphaca, Moghania vestita, mucuna capitata, Trigonella emodi, Vigna capensis, V. radiata var. sublobata, V. umbellata.*

Fruits	*Elaeagnus hortensis, Ficus palmate, fragaria indica, Morus* spp., *Prunus acuminata, P. cerasoides, P. cornuta, P. napalensis, P. prostrata, P. tomentasa, Pyrus buccata, P. communis, P. kumaoni, P. pashia, Ribes glaciale, R. nigrum, Rubus ellipticus, R.fruticosus, R. lanatus, R. lasiocarpus, R. molluccanus, Zizyphus vulgaris.*
Vegetables	*Abelmoschus manihot (tetraphyllus forms), cucumis hardwickii, C. callous, Luffa echinata, L. graveolens, Solanum incanum, S. indicum, Trichosanthes multiloba, T. himalensis.*
Oilseeds	*Lapidium capitatum, L. latifolium, L. draba, L. Ruderale.*
Fibres	*Linum perenne.*
Spices and Condiments	*Allium rubellum, A. schoenoprasum, A. tuberosum* and other species, *Carum bulbocastinum.*
Miscellaneous	*Saccharum filifolium, Miscanthus nepalensis.*
B. Eastern and North-Eastern Himalayas	
Cereals and Millets	*Digitaria cruciata, Hordeum agricrithon.*
Legumes	*Atylosia barbata, A. scarabaeoides, A. villosa, Canavalia ensiformis, Mucuna bractearata, vigna umbellate, V. radiata var. sublobata, V. pilosa.*
Fruits	*Fragaria indica, Morus spp., Myrica esculenta, Prunus acuminata, P. cornuta, P. jenkinsii, P.nepalensis, Pyrus pashia, Ribes glaciale, Rubus lineatus, R. ellipticus, R. lasiocarpus, R. molluccanus, R. reticulates, Citrus assamensis, C. ichangensis, C. indica, D. hookeriana, Eriobotya angustifolia, Mangifera sylvatica, Musa balbisiana complex, M. manii, M. nagensium, M. sikkimensis, M. supreba, M. velutina.*
Vegetables	*Abelmoschus manihot (pungens forms), Luffa graveolens, Neoluffa sikkimensis, Cucumis hystrix, C. callosus, Momordica dioica, M. cochinchinesis, M. macrophyllata, M. subangulata, Trichosanthes cucumerina, T. dioica, T. dioaelosperma, T. khasiana, T. ovata, T. trumcata, Solanum indicum and tubertypes, Allocasia macrorhiza, Amorphophallus bulbifer, Colocasia esculenta, Dioscorea alata, moghania vestica, Vigna capensis.*
Oilseeds	*Brassica trilocularis* types.
Fibres	*Gossypium aboreum* (primitive types).
Spices and Condiments	*Allium tuberosum, A. sublobatum, Curcuma zedoaria, Alpinia galanga, A. speciosa, Amomum aromaticum, Cucuma amoda, Piper peepuloides.*
Miscellaneous	*Saccharum langisetosum, S. sikkimensis, S. ravennae, Erianchus spp., Miscanthus nudipus, M. nepalensis, M. taylorii, Naranga fallax, Camellia spp.*

Source: Arora and Nayar (1984)

World's crop plant genetic resources can be more or less assigned to the specific centers of diversity identified by Vavilov (1951) based on varietal diversity, homologous variation, endemism, dominant allele frequencies, and disease resistance.

India has eight sub-centers, namely, i) Western Himalayas, ii) Eastern Himalayas, iii) North Eastern Region, iv) Gangetic Plains, v) Indus Plains, vi) Western Ghats, vii) Eastern Ghats, and viii) Andaman and Nicobar (Arora and Nayar 1984). The first three of these fall in the Himalayan Region. There are around 399 species of wild relatives of crop plants belonging to 90 genera in the Western, Eastern, and North-Eastern Himalayas.

Summary

Agrobiodiversity encompasses agroecosystems, species, and genetic diversity in cultivated plants, domesticated animals, and microorganisms used in various processes at home and industrial processes. A subset of nature's biodiversity, agrobiodiversity is critical for food security for the entire humanity as well as for food sovereignty of a country. Agrobiodiversity conservation and enhancement is an imperative of our times.

Global agriculture has undergone three phases in its entire history and has now ushered in the fourth phase. These are: i) Primitive agriculture, ii) Traditional agriculture, iii) Green Revolution agriculture, and iv) LPG agriculture, or Gene Revolution agriculture. Primitive agriculture began in the era of the dawn of civilization when man began tilling the land for the first time. This type of agriculture was not linked with the market and represented an ideal system of food sovereignty and food security. Traditional agriculture had a larger area under cultivated land but had strong linkages with uncultivated (forest/ grassland/ rangeland) areas. This type of agriculture was not strongly linked with the market system. Green Revolution agriculture is a set of research technology transfer initiatives that essentially included: i) high-yielding varieties (HYVs) of cereals – especially dwarf wheat and rice varieties with narrow straw-grain ratios; ii) chemical fertilizers, iii) pesticides, iv) frequent irrigation, and v) new methods of cultivation, including mechanization and monocultures. This agriculture had strong links with the market. The liberalization, privatization, and globalization (LPG) agriculture is controlled by big biotech companies, and foods are governed by a market system. Anthropocentric biodiversity management from ancient agriculture to LPG agriculture has registered a staggering decrease in agrobiodiversity with a corresponding increase in agricultural vulnerability and unsustainability.

Agrobiodiversity being used in cultivation practices largely varies according to i) geographical, or agroclimatic zone; ii) people's religion, culture, and traditions; and iii) market values and demands. The broader the base of the biodiversity the higher the level of food security and food sovereignty. About 75 percent of the food humans consume comes from just 12 plants and five animal sources, with about 60 percent of the calories in the entire human diet coming from three crops – wheat, rice, and corn. According to FAO, 51 percent of the calories consumed by the world's population come from wheat, rice, and corn alone. There has been a staggering decline in many species of wildlife in the last 50 years. Modern agricultural systems have pushed wildlife to an end. If the insect species continue to disappear at this rate, millions of plants will become extinct from the earth and agricultural production, and consequently food security, will be in jeopardy.

The inverted ecological pyramid indicates that the food production capacity of the agroecosystems is almost squeezed by intensive cultivation management and induced ecological vulnerability. The upright food pyramid indicates the state of sustainability, but this state is only indicative of the state just before the ecological collapse. A pyramid of agriculture reveals that the primitive type of agriculture lies at the base of the pyramid and, therefore, is the most sustainable, which is due to the very high level of biodiversity and agrobiodiversity that this agriculture harbors. LPG agriculture lying towards the tip of the pyramid, on the other hand, is most unsustainable. Ecological affluence ensures plentiful foods obtainable from innumerable plants, cultivated as well as uncultivated. Ecological affluence, on the other hand, is ensured by pursuing, following, and sustaining the most appropriate food production systems. And sustainable future is ensured by a healthy and ameliorating relationship between the two.

Mountain areas are altogether distinguishable from those of the plains. Mountains exhibit six common characteristics, viz., marginality, limited accessibility, fragility, diversity of heterogeneity, ecological niches or comparative advantage, and human adaptation mechanisms. The first three characteristics are the negative attributes of the mountains, whereas the last three are the positive attributes. An agroecosystem is an arbitrarily defined unit of nature with distinct energy flows and chemical cycling, involving uncultivated land areas (forests, grasslands/ grazing lands), cultivated land, and livestock in organic linkages with each other, woven into a complex unitary whole, functionally oriented to produce foods and other life-supporting products – such as feed, fiber, fuel, fertilizers, etc. – while maintaining ecological equilibrium. Uncultivated land area – forests, grasslands, grazing lands – is a core component and the whole biodiversity conserved in this component nourishes the cultivated land through a constant flow of nutrients. Cultivated land is managed through ethno-engineering often by clearing original forests. Livestock are the dynamic component of an agroecosystem; nutrient flows from uncultivated land are mediated by livestock. A community of farmers living in a cluster of houses called a village manages an agroecosystem.

The Himalayan Mountain Region has provided an appropriate environment for several varieties of cereals, pulses, fruits, oil-yielding plants, spices, tuberous vegetables, and sugar-yielding plants and their close relatives that number 155. World's crop plant genetic resources can be more or less assigned to the specific centers of diversity identified by Vavilov based on varietal diversity, homologous variation, endemism, dominant allele frequencies, and disease resistance. India has eight sub-centers, namely, i) Western Himalayas, ii) Eastern Himalayas, iii) North Eastern Region, iv) Gangetic Plains, v) Indus Plains, vi) Western Ghats, vii) Eastern Ghats, and viii) Andaman and Nicobar. The first three of these fall in the Himalayan Region. There are around 399 species of wild relatives of crop plants belonging to 90 genera in the Western, Eastern, and North-Eastern Himalayas.

Exercises

A. Multiple Choice Questions

1. Which of the following organisms is a photoautotroph?

   a. Algae b. Blue-green algae

   c. A mango tree d. All of the above

2. Which of the following organisms is a heterotroph?

   a. Human being b. Whale

   c. Locust d. All of the above

3. Which of the following organisms follows carnivory?

   a. A cat b. A leopard.

   c. A lizard d. All of the above

4. Which of the following organisms are dependent on chemolithoautotrophy/ chemoautotrophy?

   a. Nitrogen-fixing bacteria b. Blue-green algae

   c. Giant tubeworm d. All of the above

5. Which of the following Kingdoms of organisms can be photoautotrophs, chemoautotrophs as well as heterotrophs?
   a. Prokaryotes (bacteria, archaea)   b. Protists
   c. Fungi   d. Plants
6. Which of the following Kingdoms could include photoautotrophs as well as heterotrophs?
   a. Plants   b. Fungi
   c. Animals   d. None of the above
7. Agrobiodiversity refers to
   a. Domesticated plants and animals   b. All species in a natural forest
   c. All species in an open ocean   d. All insects
8. Which of the following could be an example of agrobiodiversity?
   a. A corn plant   b. A shark
   c. A crocodile   d. Earthworms
9. Which type of agriculture hosted the highest level of agrobiodiversity?
   a. Green Revolution   b. Gene Revolution
   c. Primitive   d. Traditional
10. Which of the following types of agriculture hugely depends on chemical fertilizers and pesticides?
   a. Primitive   b. Traditional
   c. Green Revolution   d. None of the above
11. If the history of agriculture is plotted with a pyramid in the background, what phase of agriculture would lie at the base of an agriculture pyramid?
   a. LPG   b. Green Revolution
   c. Primitive   d. Traditional
12. What portion of a pyramid of agriculture is occupied by LPG agriculture?
   a. Lowest portion   b. Middle portion
   c. Tip   d. All of the above

13. What era of agriculture has represented the highest degree of sustainability?
    a. Traditional agriculture
    b. Primitive agriculture
    c. Green Revolution-type agriculture
    d. Gene revolution-type agriculture
14. What type of agriculture is based on biotechnology?
    a. Green Revolution
    b. Gene revolution
    c. Traditional
    d. Primitive
15. In the Bt crops, what does Bt represent?
    a. A protozoa
    b. A bacterium
    c. A fungus
    d. An algae
16. LPG agriculture is also referred to as
    a. Green Revolution
    b. Gene Revolution
    c. Food Revolution
    d. Organic Revolution
17. Who did for the first time use the term 'Green Revolution'?
    a. William S. Gaud
    b. Norman E. Borlaug
    c. M. S. Swaminathan
    d. Vandana Shiva
18. Which of the following mountain agroecosystem specificities is a positive attribute?
    a. Fragility
    b. Ecological niches
    c. Marginality
    d. Limited accessibility
19. Which set of mountain agroecosystem characteristics reflects negative attributes?
    a. Fragility, Marginality, Diversity/ Heterogeneity
    b. Limited accessibility, Ecological niches, Marginality
    c. Marginality, Fragility, Limited Accessibility
    d. Ecological niches, Diversity/ Heterogeneity, Human adaptation mechanisms

20. According to the study of Arora and Nayar, the number of species of wild relatives of crop plants belonging to 90 genera in the Western, Eastern, and North-Eastern Himalayas is

a. 1000 b. 595

c. 499 d. 399

Answers

1. **d**, 2. d, 3. **d**, 4. **d**, 5. **a**, 6. **a**, 7. **a**, 8. **a**, 9. **c**, 10. **c**, 11. **c**, 12. **c**, 13. **b**, 14. **b**, 15. **b**, 16. **b**, 17. **a**, 18. **b**, 19. **c**, 20. **d**

B. Short-Answer Questions

1. What are the differences between autotrophs and heterotrophs?
2. What do you mean by chemolithoautotrophs?
3. Which kingdom of living species embraces photoautotrophs, chemoautotrophs, and heterotrophs? Give some examples.
4. What is agrobiodiversity? Write citing examples.
5. Mention some differences between biodiversity and agrobiodiversity.
6. Name the four eras representing different agricultural systems.
7. How did primitive agriculture harbor an extreme level of biodiversity/ agrobiodiversity?
8. Which type of agricultural system does represent a narrow base of agrobiodiversity?
9. Name the components of an agroecosystem such as in the mountains.
10. What do you mean by marginality as a feature of mountain ecosystems?

C. Long-Answer Questions

1. What is an agroecosystem? Why does our agriculture need to be based on an agroecosystem rather than exclusively on croplands?
2. How has the biodiversity-agrobiodiversity pattern changed over the history of agriculture development? Discuss with reasoning.
3. Discuss agrobiodiversity and food production relationships.
4. What are the major differences between mountain and mainstream agriculture? Explain with a flow diagram the essential features of a mountain agroecosystem.

5. Write short notes on:
    a. Pyramid of Agriculture
    b. Livestock in a Farming System/ Agroecosystem
    c. Wild Relatives of Crop Plants in the Himalayas

References

Arora, R. K. and Nayar, E. R. 1984. Wild Relatives of Crop Plants in India. New Delhi: National Bureau of Plant Genetic Resources (NBPGR).

Jodha, N. S. 1990. Mountain Agriculture: Search for Sustainability. Mountain Farming Systems Discussion Paper No. 2. Kathmandu: ICIMOD.

Jodha, N. S. 1993. A Framework for integrated mountain development. In Singh, V. (ed.) Eco-Crisis in the Himalayas: Causes, Consequences and Way Out. Dehradun: IBD.

Jodha, N. S., Banskota, M. and Partap, T. 1992. Sustainable Mountain Agriculture. New Delhi: Oxford & IBH Co.

Knorr and WWF. 2019. Future 50 Foods for Healthier People and a Healthier Planet. 60 pp. Link: https://www.knorr.com/content/dam/unilever/knorr_world/global/online_comms_/knorr_future_50_report_online_final_version-1539191.pdf.

Reijntjes, C., Haverkort, B. and water-Bayer, A. 1992. Farming for the Future: An Introduction to Low-External-Input and Sustainable Agriculture. London: MacMillan Press. 250pp.

Singh, Vir. 2005. Agrobiodiversity, Sustainability and Food Security in the Himalayan Mountains: An Uttaranchal Perspective. Gorakhpur: Gorakhpur Environmental Action Group, 50 pp.

Singh, Vir. 2014. The Diversity of Life: A Living Terrestrial Wonder of Cosmic Light. GB Pant University of Agriculture and Technology, Pantnagar, India, 22pp.

Singh, Vir. 2019. Fertilizing the Universe: A New Chapter of Unfolding Evolution. London: Cambridge Scholars Publishing. 285pp. (ISBN (10): 1-5275-2301-2 ISBN (13): 978-1-5275-2301-2).

Singh, Vir. and Tulachan, P. M. 2002. Marginal Farming in Mountain Areas. Asian Agri-History, 6(3): 269-280.

Singh, Vir, Shiva, V. and Bhatt, V. K. 2014. Agroecology: Principles and Operationalisation of Sustainable Mountain Agriculture. New Delhi: Navdanya. 64p+viii.

Singh, Vir. 2018. Food security amidst the climate change scenario: Perspectives, issues and opportunities in mountain agriculture. Nutri Food Sci Int J, 6(3): 555686. doi: 10.19080/NFSIJ.2018.06.555686.

Singh, Vir. 2022. Three crops rule the world: What it means for the planet's wildlife? State of India's Environment (2022)

Vavilov, N. I. 1951. The origin, variation, immunity and breeding of cultivated plants. Chronica Botanica 13: 1-364.

Web Resources

http://cpicfinance.com/blog-biodiversity-credits-a-turning-point-for-nature/?fbclid=IwAR25_tZE4yTh4IMu19cQ-uiuxYDQjmQDiSDqncUK9zrtEg_dqwAEcOMr5wo

https://www.downtoearth.org.in/news/food/three-crops-rule-the-world-what-it-means-for-the-planet-s-wildlife-81781

https://www.knorr.com/us/en/future-50/why-it-matters/future-50-foods-for-a-healthier-planet.html

https://www.worldwildlife.org/projects/plowprint-report

3

Biodiversity of the Pedosphere

All of nature's biodiversity is not visible. Biodiversity can also not be studied and/ or understood by looking at and analyzing it with the naked eye. A great proportion of the planet's biodiversity flourishes in the ecosystems not directly accessible to us and the organisms in the sizes our eyes cannot record. The soil biodiversity is more interesting and more varied than we have understood and appreciated. The biodiversity in the soil – together with the physical environment – constitutes what we can call the pedosphere. Biodiversity in the pedosphere is distinguishable from that of the biodiversity above the soil surface. One is nonplussed by an understanding of some of the unbelievable features and functions of the pedosphere biodiversity.

The Convention on Biodiversity (CBD) held on the occasion of the historical Earth Summit in Rio de Janeiro in 1992 underlined the significance of biodiversity conservation for various tangible and non-tangible benefits as well as one of the key factors for sustainability. However, the conservation of soil biodiversity for realizing sustainability in terrestrial ecosystems was not given the required emphasis on this occasion. However, upon ushering in the new millennium, due importance to soil community conservation was given in international environmental policies, notably in EU Soil Thematic Strategy (2006), the Biodiversity Action Plan for Agriculture (EU 2001), the Kyiv Resolution on Biodiversity (EU/ ECU 2003) (Menta 2012).

The Pedosphere

The pedosphere, the outermost layer of the Earth composed of soil and subject to soil formation processes, is an integral component of the biosphere, lying on the interface of the atmosphere, hydrosphere, and lithosphere, and much more complex in its physicochemical and biological characteristics and functioning and quite distinguishable from that of the above-soil surface terrestrial and aquatic ecosystems. The pedosphere has its peculiar environment supporting and sustaining an amazing variety of organisms organized in trophic levels of auto- and heterotrophs and making up a variety of communities of interdependent organisms in a variety of soil ecosystems.

The physical or abiotic environment of the pedosphere comprises gases, water/ moisture, and inorganic and organic substances. The inorganic substances include a variety of minerals – metallic and non-metallic elements, including all the nutrients (macro- and micronutrients) necessary for the growth and maintenance of all soil organisms and plants growing on the soil, and inorganic compounds and mixtures. Soil organic matter (SOM), i.e., the organic matter consisting of three primary parts – including small (fresh) plant residues, small living soil organisms, decomposing (active) organic matter, and stable organic matter (humus) – on the whole, contains more than three times as much carbon as either the atmosphere or terrestrial vegetation (Negi et al. 2018). Organic substances include dead remains of plant and animal substances and compounds like carbohydrates, proteins, fats, oils, nucleic acids, etc. along with humus, the stable organic matter.

The biotic component of the pedosphere often comprises a rich diversity of living species, their genotypes, and communities. Dominated by microorganisms, the pedosphere also embraces the flora and fauna belonging to higher living forms. The organisms belong to all the five kingdoms of life – monera, protista, fungi, plants, and animals. The above-soil surface life is also largely determined by the life in the pedosphere. So important is the pedosphere for the entire biosphere!

Pedodiversity

Soil is the largest ecosystem on Earth, and diversity of the soils, or pedodiversity, and biodiversity in the soil ecosystems, or pedobiodiversity, constitute one of the unique characteristics of the soil ecosystem, or pedoecosystem, in particular and of nature's biodiversity in general. Soils of the Earth vary according to their geographical positioning/ climatic region, physical structures, chemical composition, and biological features (Singh 2020). The soil environment is influenced seminally due to vegetation cover. Since vegetation types and cover in our world are largely determined by human activities, anthropogenic factors remain dominant to eventually influence soil ecology.

Based on their structure and composition, the planet's soils have been broadly classified into six main types: sandy, silty, clay, loamy, and chalky, which are made up of the three major soil types: sandy, silty, and clay. Of all the soil types, sandy soils have the largest particle size, and clay soils have the smallest. The soil particle sizes determine the space sizes formed within a soil ecosystem and thus water drainage rates/ water and nutrient holding qualities of the soils, and as a consequence, the quality of the soil environment necessary for soil biodiversity and plant growth status. Among all the soil types, loam is regarded as "perfect" with an appropriate proportion of the three major soil types – sand (40%), silt (40%), and clay (20%) – and a high proportion of carbon. This soil is also considered to be the most ideal one for crop cultivation. Salient physicochemical features of different soil types are presented in Fig. 3.1.

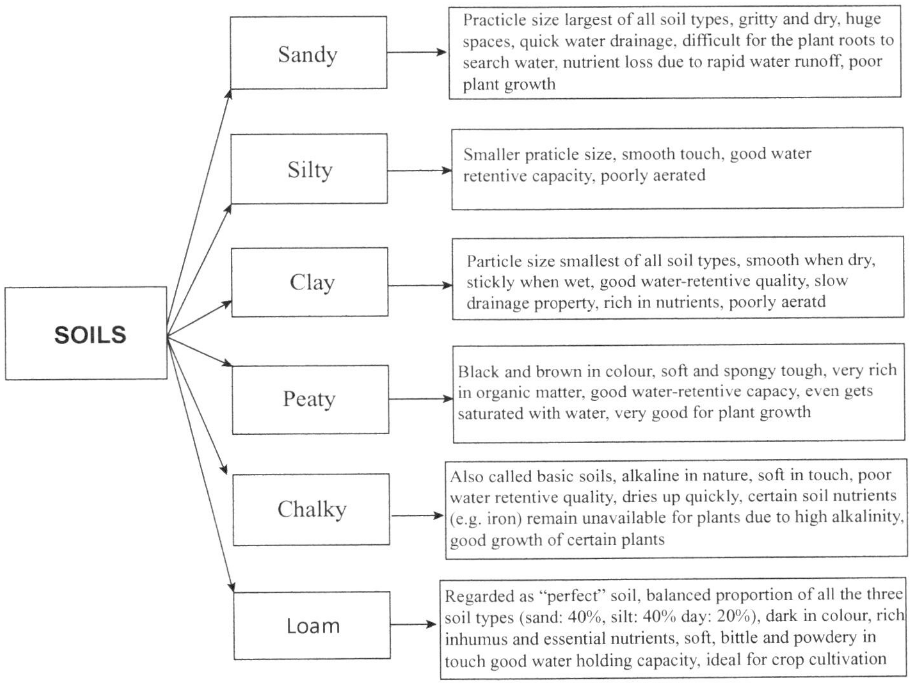

Fig. 3.1: The Pedodiversity: Major soil types of the world and their salient characteristics

An "Invisible World" of Biodiversity

Although invisible, the pedosphere is home to a unique and richest "world" of biodiversity. There is, in fact, an amazing world beneath our feet (Singh 2019). Soils host a quarter of our planet's biodiversity (FAO 2015). Soil is the foundation of terrestrial life. A scenario of the diversity of life within the soil habitats of the world that baffles the human mind is as follows (FAO 2015, Singh 2019, 2020):

- More than 1000 species of invertebrates may be found in a single square meter of forest soil.
- Planet's many terrestrial insect species inhabit soil for at least some stage of their life cycle.

- One gram of soil might harbor millions of individuals and thousands of bacteria species.
- Healthy soil may inhabit thousands of species of bacteria and actinomycetes, hundreds of species of fungi, tens of nematode species, 50-100 species of insects, 20-30 species of mites, several species of earthworms, and several species of vertebrate animals.
- Soil contains the organisms with the largest area; e.g., a single colony of the honey fungus (*Armillaria ostoyae*) may cover approximately 9.0 square kilometers.

The "invisible world" beneath our feet ecstatically blossoms with the biodiversity which measures more than that of the biodiversity of the "visible world" in every respect: number of species and their genotypes, density, distribution, and probably biomass. However, this "invisible" biodiversity depends on the ecological state of the soil ecosystems. Living soil destroyed by desertification and/ or damaged by pollutants is likely to harbor a very small measure of biodiversity. The biodiversity of the "visible world" is rooted in the biodiversity of the "invisible world". In another world, the state of biodiversity of the terrestrial ecosystems we witness with our own eyes depends on the biodiversity of the soil ecosystems – the pedosphere.

Biodiversity of the pedosphere is of critical importance in playing a crucial role in chemosynthesis, photosynthesis, nitrogen fixation, sedimentary cycles, and in creating a partial chemosynthesis-based soil community, various trophic levels, food chains, and food webs (Fig. 3.2). The primary productivity of a soil ecosystem is generated primarily via chemosynthesis maintained mainly by soil bacteria and secondarily via photosynthesis by cyanobacteria and algae, and both the autotrophic mechanisms become the basis of secondary productivity.

Photosynthesis in the above-soil terrestrial ecosystems also plays a key role by infusing carbon in the pedosphere, which also feeds a variety of detritivores in addition to adding to the fertility status of the soil. Most soil biodiversity of heterotrophs is supported by the organic matter emanating from photosynthesis in the terrestrial ecosystems.

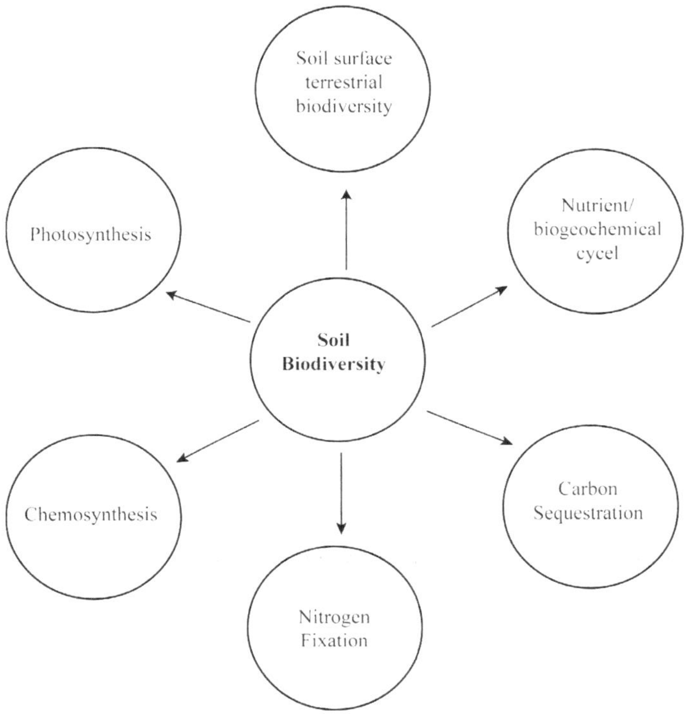

Fig. 3.2: Vital role of soil biodiversity

Some microorganisms in a soil ecosystem, such as blue-green algae or cyanobacteria, play a dual role: nitrogen fixation and photosynthesis. Photosynthesis performed by these organisms also contributes to the organic matter in the soil augmenting the fertility of the ecosystem.

Comprising bacteria, actinomycetes, fungi, cyanobacteria, and algae, the soil microflora represents a very high level of diversity and is more vibrant, varied, and more stable than the soil fauna that includes, protozoa, nematodes, Collembola or springtails (small wingless arthropods also considered to be the most abundant macroscopic organisms on Earth), and acarids which tend to move about upon exhaustion of their foods. A broad classification of the wide array of organisms based on taxonomic groups and their sizes that inhabit diverse soils is shown in Fig. 3.3.

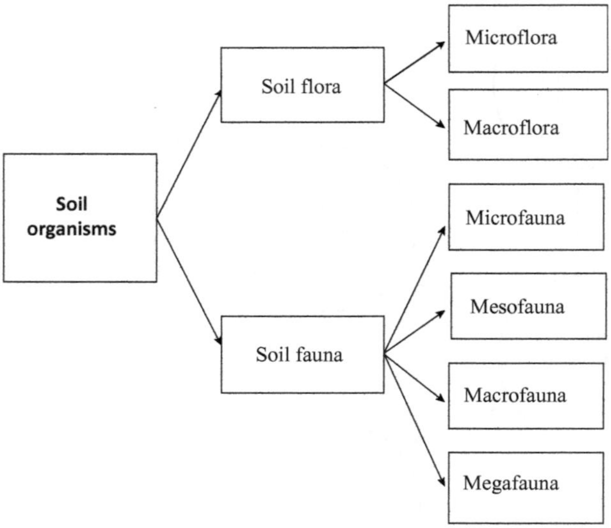

Fig. 3.3: Life in the soil: broad categories of the living organisms in the soil

The soil organisms can be classified into three groups, viz., i) micro-flora and micro-fauna, ii) mesofauna or meiofauna, and iii) macro-fauna. The flora of meso- and macro sizes inhabit land surfaces with roots in the soil, therefore they are not counted as soil organisms.

Soil Flora

Soil flora is an integral part of the organic component of soil organic matter (SOM). The pedosphere is a rich repository of microflora constituting a major biotic component of the soil.

Microflora

Bacteria are dominant in the entire soil microflora making up about 90 percent of the total microflora population. Actinomycetes constitute about 9 percent and fungi and algae together represent nearly 1 percent microflora population. All the soil properties are largely influenced by the microflora, especially by bacterial populations. Microflora, in essence, is the key biotic factor determining the health of the pedosphere. Most of the energy in the pedosphere flows through the microflora.

Neutral soils well fed by the organic matter are appropriate for the bacterial populations to flourish. Bacteria are among the organisms which are both autotrophs and heterotrophs. The autotrophic bacteria use inorganic molecules as their energy source and atmospheric carbon dioxide as a carbon source for nourishing themselves through the process what is known as chemosynthesis. The chemosynthetic bacteria, thus, are also designated as chemoautotrophs.

The chemoautotrophic bacteria for their energy source depend on a variety of inorganic substances, and they are categorized as nitrifying bacteria, sulfur bacteria, iron bacteria, manganese bacteria, methane bacteria, carbon monoxide bacteria, etc. Soils also harbor photosynthetic organisms, such as algae and blue-green algae, the cyanobacteria. Such autotrophic organisms are also designated as photoautotrophs.

The majority of the bacteria in the soil environment, however, are heterotrophs. The heterotrophic bacteria depend on organic matter as a source of energy and carbon and are mainly concerned with decomposition processes in the soil leading to mineralization. Considerable amounts of essential minerals used by plants as nutrients are released during the process of decomposition enacted by bacteria.

Specific soil biodiversity is composed of nitrogen-fixing bacteria, actinomycetes, and cyanobacteria (blue-green algae). These soil organisms are involved in the process, namely, biological nitrogen fixation (BNF). These include:

1. Free-living bacteria or symbiotic bacteria are: a) aerobic bacteria and anaerobic bacteria (e.g., *Clostridium* and *Azotobacter* respectively, inhabiting soils where they play their role in N_2 fixation), and b) blue-green algae, or cyanobacteria (photosynthetic autotrophs inhabiting waterlogged areas and wet soil, e.g. Anabaena, Nostoc, Scytonema algae, etc.).
2. Symbiotic bacteria with several fungi and nodule-forming bacteria as the two major sources of nitrogen fixation: a) Many species of Actinomycetes with the ability to fix atmospheric N_2 establish a mycorrhizal association – ecotropic and endotrophic – with the roots of certain plants, such as *Pinus*, and *Casuarina*, and b) nodule forming *Rhizobium* species.

Most of the fungi occupy acidic soils. The fungi by their nature are heterotrophs and to fulfill their nourishment they prevail as parasites, saprophytes, and symbiotics. Parasitic fungi infect roots and other parts of the plants and become the sole cause of several plant diseases, such as root rot, wilts, blights, rusts, and smuts. Some wilt-forming parasitic fungi release chemical substances which are extremely harmful to the infected plants. For example, *Fusarium oxysporum* f.sp. *lini*, a fungus responsible for producing wilt in flax, produces hydrogen cyanide (HCN), while *Fusarium udum*, causing wilt disease in pigeon pea (*Cajanus cajan*) in tropical regions, releases fusaric acid in the roots of the plants they infect. Certain parasitic fungi, however, also secrete substances that function as growth stimulators. For instance, *Fusarium* spp. are known to produce a hormone, gibberellic acid ($C_{19}H_{22}O_6$ – also called gibberellins A3, GA, and GA3), which benefits the host plants by ameliorating their growth.

Saprophytic fungi derive carbon and energy from the organic matter they perpetuate on. In the decomposition process, the saprophytic fungi break down cellulose, lignin, starch, sugars, proteins, etc. into their constituent components.

Many fungal species inhabiting various soil environments have shown antibiotic properties. These antibiotics are of high socioeconomic value. Some of the fungal species isolated from soils have exhibited some kind of antimicrobial action. These are: *Absidia corymbifera, Alternaria alternata, Aspergillus flavus, Aspergillus fumigants, Aspergillus niger, Cladosporium herbarum, Cuvularia lunata, Penicillium* sp., *Rhizopus stolonifer,* and *Trichoderma viride*. Such soil fungi may be harnessed for their antibiotic potential by the pharmaceutical industry.

Macroflora

Macroflora is not confined only to the soil ecosystem. It is also a part of the ecosystem above the soil surface. However, a significant proportion of the macroflora lies within the soil ecosystem and the overall flora leaves a phenomenal impact on soil life. Macroflora inhabiting soil ecosystems can be categorized into five ecological classes based on soil characteristics on which they thrive:

1. **Oxylophytes:** plants that prefer or are restricted to acid soil; for example, bog mosses, various heath shrubs, dwarf birch, sundew, *Rumex* spp., *Rhododendron*, etc.
2. **Halophytes:** plants occurring in saline soils; for example, *Salsola foetida*, *Salicoraia* spp., *Suaeda fruticosa*, etc.
3. **Psammophytes:** plants growing on sand/ sandy soils; for example, *Agriophyllum squarrosum* (sand rice), *Haloxylon ammodendron*, *Linaria arenaria*, etc.
4. **Lithophytes:** plants growing on rock surfaces; for example, several orchids (*Dendrobium* and *Paphiopedilum*, etc.), bromedias, ferns, algae, liverworts, and also many species of carnivorous plants, such as pitcher plants *Nepenthes campanulata* and *Heliamphora exappendiculata*, etc.
5. **Chasmophytes:** plants that grow on rock crevices or anywhere on the rocks where detritus gets accumulated; for example, *Asarina procumbens*, and several other plants that can also grow on rocks.

Soil Fauna

Soil fauna is a very crucial and interesting biotic component of a soil ecosystem responsible for several functions. Many soil animals are considered key indicators of soil quality.

Microfauna

Animals dwelling in the soil environment which bear sizes of 20 μm to 200 μm (< 100 μm, according to some authors) are placed in the category of soil microfauna. Protozoa, small-sized mites, nematodes, rotifers, tardigrades (eight-legged segmented micro-animals), and copepod crustaceans belong to the microfauna category of soil organisms. The organisms in this group live at varying depths of the soil habitat. For example, protozoans, like amoeba, ciliates, and zoomastigina flagellates live near the soil surface, whereas the testate forms, such as *Thecamoeba*, *Euglypha*, and *Difflugia* enjoy wider vertical distribution.

Nematodes in the soil habitats occur in plentiful numbers. For example, the nematodes, like *Rhabditis*, *Diplogaster*, *Mononchus*, *Xiphinema*, *Heterodera*, *Criconemoides*, and *Aphelenchoides* prevail as plentifully as 1 to 3 million m^{-2} in raw humus soils to 20 million m^{-2} in grassland soils.

Mesofauna

Soil mesofauna species include animals with sizes ranging from 200 μ to 1 cm (0.1-2 mm size, according to some studies). Micro-arthropods Acarina (mites) and Collembola (springtails) make the important members of this group that also includes the larger nematodes, rotifers, and tardigrades, along with the majority of isopods, Arachnida, Chelognathi, Opiliones, Enchytraeidae insect larvae and small millipedes, isopods and mollusks.

Populations and density of the species belonging to soil mesofauna vary as per the soil composition, especially the proportions of organic matter and moisture. Management practices relating to soils also greatly influence mesofauna.

Macrofauna

Soil animals measuring more than 1 cm in body size fall under the category of macrofauna. According to Gongalsky (2021), the body size of the soil animals designated as macrofauna is between 1-2 mm and 20-30 mm. In some ecosystems, the macrofauna may account for most of the total soil animal biomass and substantially contributes to soil food-web functioning (Gongalsky 2021). Major types of soil animals include those belonging to Lumbricidae (a family of earthworms), Mollusca (the second largest phylum of invertebrate animals after the Arthropoda), and also large-sized chilopods, arachnids, insects, and fossorial vertebrates (amphibians, reptiles, birds, and mammals).

Earthworms, the Annelids, in the soil have got a special reputation as being the key visible indicators of soil fertility status because of their role in breaking down litter fragments, organic matter decomposition, mixing up of organic matter well with soil minerals, and increasing aeration rates in the soil necessary for plant growth. The common Indian Annelid species dwelling in the soil are: *Pheretima* (found mainly in New Guinea and parts of South Asia), *Megascolex* (found in the soils in Madagascar, Australia, New Zealand, North America, and South East Asia), *Octochaetus* (native to Australia, New Zealand and Malaysia), *Drawida* (commonly found in India and China), and *Moniligaster* (generally restricted to the Western Ghats mountain regions of southern Peninsular India) (Narayanan et al. 2022). Numerical dominance of the earthworms is found in alkaline soils with high organic matter and adequate amounts of moisture. In acidic soils, earthworms exhibit only a sparse presence.

Scolopendra (species of centipedes) and *Lithobius* (stone centipedes) among chilopods and *Gryllotalpa* (European mole cricket widely distributed in Europe) and *Forticula* (an omnivorous insect, *F. auricularia*, being the common earwig or European earwig) among insects are well adapted to fossorial life in soil habitats. Soils rich in lime are more appropriate for the land snails.

Megafauna

Soil vertebrates are adapted to both soil habitats and above-soil terrestrial habitats, and some of them even to aquatic habitats. In the soil, the vertebrates usually dwell in fossorial or burrowing "lifestyles". The common soil vertebrates among Amphibians include *Breviceps* (species commonly known as rain frogs or short-headed frogs), and *Ichthyophis* (a genus of caecilians—limbless amphibians, sometimes called the Asian caecilians); among Reptiles, they include *Sphenodon* (or hatteria, the last survivor of the reptilian order *Rhynchocephalia* believed to have evolved in the early Mesozoic era), *Uromastyx* (a genus of African and Asian agamid lizard), limbless lizards, and snakes; among Avian species, it is burrowing owl; and the common mammals include kangaroo-rats, ground squirrels, otter, moles, rodents, badgers, hedgehog (a spiny mammal), etc.

Soil Biodiversity Erosion

Soil biodiversity varies according to many natural factors, such as soil type, soil composition, topography, climate, etc. Any natural factor(s) leading to soil degradation, soil erosion, and eventually desertification is/ are detrimental to soil biodiversity. The most striking and critical factors of soil biodiversity crises are the anthropogenic factors (Fig. 3.4). Land use change has been one of the major causes of soil biodiversity loss. Transformation of forest areas, grasslands, rangelands, etc. into cultivated lands has invited multiple environmental and ecological crises which are ultimately rooted in the crises of soil ecosystems.

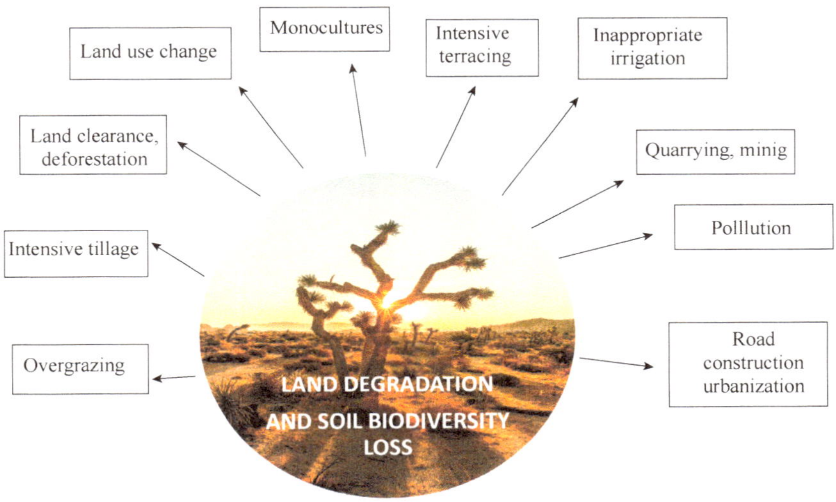

Fig. 3.4: Anthropogenic factors contributing to land degradation and consequent soil biodiversity depletion

According to an estimate, a large chunk of agricultural land – approximately 40% – is degraded (Sample 2007). The Special IPCC Report (IPCC 2019) further makes us cautious about land degradation:

- About a quarter of the Earth's ice-free land area is subject to human-induced degradation; and
- Soil erosion from agricultural fields is estimated to be currently 10 to 20 (no-tillage) to more than 100 times (conventional tillage) higher than the soil formation rate.

A land area covered with dense perennial vegetation, e.g. a natural forest, builds up a soil ecosystem rich in biodiversity. The forest soil is constantly fed with organic matter by photosynthesizers. The vegetation cover also maintains soil moisture necessary for soil life. Vertical nutrient cycles the deep-rooted trees play a vital role in contribute to maintaining the rich treasure of all plant nutrients in the soil, and vegetation also checks soil erosion.

When an area is deforested and converted into cultivated lands, processes leading to soil erosion are triggered (Fig. 3.5). Cultivation practices involving intensive tillage, frequent irrigation, monocultures of crop annuals, and frequent applications of chemical fertilizers and pesticides open land area to unabated soil erosion, soil biodiversity annihilation and eventually to desertification processes. Mismanagement of the soil in an area further escalates the processes that on a long-term basis end up in an unproductive desert. A natural forest evolved up to its ecological climax through succession blossoms with the highest degree of biodiversity. Deforestation followed by unabated land degradation eventually leads to a state of disclimax via the process of desertification that represents minimum or zero biodiversity (Fig. 3.6).

Fig. 3.5: Modern agriculture, often praised as the Green Revolution agriculture, is responsible for most of the soil erosion and biodiversity loss on the face of the Earth (Photo by Dylan de Jonge on Unsplash).

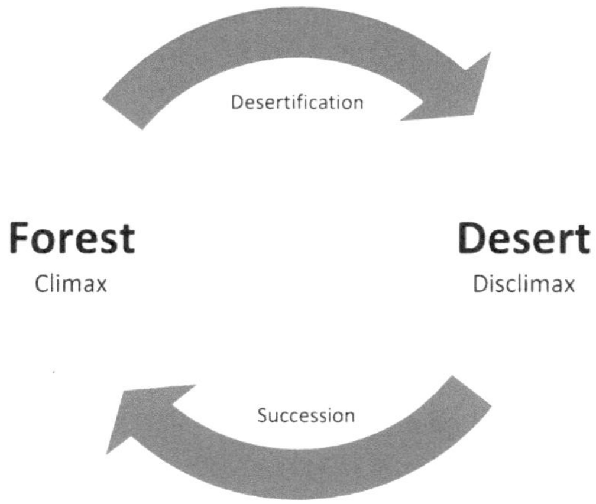

Fig. 3.6: The two 'poles' of biodiversity: Ecological climax with the highest degree of biodiversity and ecological disclimax with the collapse of biodiversity

Conserving Soil Biodiversity

Soil, home to numerous types of living organisms, undoubtedly is the most active ecosystem on Earth as well as a natural foundation of terrestrial life on Earth. Someone has rightly said that if you own soil, you own up to the sky. Soil is also a resource on which humanity's food security depends. Biodiversity in the pedosphere – thus – serves as a resource on which humanity's future depends. Soil biodiversity, in essence, is an axis of sustainability of life on the planet.

Since soil biodiversity is depleting due to human activities, its conservation can also be ensured by conservation-oriented human activities. Vegetation cover on the soil surface, especially the natural forests with trees as the dominant members of the forest community, will be a major determinant of biodiversity conservation. The soil surface biodiversity (the forest biodiversity) influences soil biodiversity and vice versa. Protection and augmentation of existing forests, afforestation on a massive scale in the deforested areas, and reclamation of degraded land employing various ecological measures are the priority for the pedosphere's diversity to restore and perpetuate, and function.

Agricultural practices which, unfortunately, are not conducive to soil conservation need a paradigm shift. Ecologically sound soil fertility management, cropping systems involving a synergistic combination of deep root and shallow root crops, greater emphasis on agroforestry and no-till farming, and reestablishing forestry-cropping relationships will greatly ameliorate soil environment and biodiversity. Agriculture development based on sustainability principles – that is, living soil, biodiversity, and cyclic nutrient flows – will prove to be most promising for the natural biodiversity to prevail within and outside the pedosphere.

Summary

The pedosphere has its peculiar environment supporting and sustaining an amazing variety of organisms organized in trophic levels of auto- and heterotrophs and making up a variety of communities of interdependent organisms in a variety of soil ecosystems. Soil is the largest ecosystem on Earth. Based on their structure and composition, the planet's soils have been broadly classified into six main types: sandy, silty, clay, loamy, and chalky, which are made up of the three major soil types: sandy, silty, and clay. Among all the soil types, loam is regarded as "perfect" with an appropriate proportion of the three major soil types – sand (40%), silt (40%), clay (20%) – and a high proportion of carbon. This soil is also considered to be the most ideal one for crop cultivation.

Soils host a quarter of our planet's biodiversity. More than 1000 species of invertebrates may be found in a single square meter of forest soil. One gram of soil might harbor millions of individuals and thousands of bacteria species. Healthy soil may inhabit thousands of species of bacteria and actinomycetes, hundreds of species of fungi, tens of nematode species, 50-100 species of insects, 20-30 species of mites, several species of earthworms, and several species of vertebrate animals. Soil contains the organisms with the largest area; e.g., a single colony of the honey fungus (*Armillaria ostoyae*) may cover approximately 9.0 square kilometers.

Biodiversity of the pedosphere is of critical importance in playing a crucial role in chemosynthesis, in photosynthesis, nitrogen fixation, sedimentary cycles, and in creating a chemosynthesis-based soil community, various trophic levels, food chains, and food webs. Some microorganisms in a soil ecosystem, such as blue-green algae or cyanobacteria, play a dual role: nitrogen fixation and photosynthesis. Photosynthesis performed by these organisms also contributes to the organic matter in the soil augmenting the fertility of the ecosystem.

Bacteria are dominant in the entire soil microflora making up about 90 percent of the total microflora population. Actinomycetes constitute about 9 percent and fungi and algae together represent nearly 1 percent microflora population. The chemoautotrophic bacteria for their energy source depend on a variety of inorganic substances, and they are categorized as nitrifying bacteria, sulfur bacteria, iron bacteria, manganese bacteria, methane bacteria, carbon monoxide bacteria, etc. Soils also harbor photosynthetic organisms, such as algae and blue-green algae, the cyanobacteria. Such autotrophic organisms are also designated as photoautotrophs. The majority of the bacteria in the soil environment, however, are heterotrophs that depend on organic matter as a source of energy and carbon. Most of the fungi occupy acidic soils. The fungi by their nature are heterotrophs and to fulfill their nourishment they prevail as parasites, saprophytes, and symbiotics. Parasitic fungi infect roots and other parts of the plants and become the sole cause of several plant diseases, such as root rot, wilts, blights, rusts, and smuts. Saprophytic fungi derive carbon and energy from the organic matter they perpetuate on. In the decomposition process, the saprophytic fungi break down cellulose, lignin, starch, sugars, proteins, etc. into their constituent components.

Macroflora inhabiting soil ecosystems can be categorized into five ecological classes based on soil characteristics on which they thrive: Oxylophytes (plants that prefer or are restricted to acid soil); Halophytes (plants occurring in saline soils); Psammophytes (plants growing on sand/ sandy soils); Lithophytes (plants growing on rock surfaces); and Chasmophytes (plants that grow on rock crevices or anywhere on the rocks where detritus gets accumulated)

Animals dwelling in the soil environment which bear sizes of 20 μ to 200 μ are placed in the category of soil microfauna. Protozoa, small-sized mites, nematodes, rotifers, tardigrades, and copepod crustaceans belong to the microfauna category of soil organisms.

Soil mesofauna species include animals with sizes ranging from 200 μ to 1 cm. Micro-arthropods Acarina (mites) and Collembola (springtails) make the important members of this group that also includes the larger nematodes, rotifers, and tardigrades, along with the majority of isopods, Arachnida, Chelognathi, Opiliones, Enchytraeidae insect larvae and small millipedes, isopods and mollusks.

Populations and density of the species belonging to soil mesofauna vary as per the soil composition, especially the proportions of organic matter and moisture. Management practices relating to soils also greatly influence mesofauna. Soil animals measuring more than 1 cm in body size fall under the category of macrofauna. Major types of soil animals include those belonging to Lumbricidae (a family of earthworms), Mollusca, and also large-sized chilopods, arachnids, insects, and fossorial vertebrates (amphibians, reptiles, birds, and mammals). Earthworms, the Annelids, in the soil have got a special reputation as being the key visible indicators of soil fertility status because of their role in breaking down litter fragments, organic matter decomposition, mixing up of organic matter with soil minerals, and increasing aeration rates in the soil necessary for plant growth. In the soil, the vertebrates usually dwell in fossorial or burrowing "lifestyles".

The most striking and critical factors of soil biodiversity crises are anthropogenic factors. Land use change has been one of the major causes of soil biodiversity loss. A land area covered with dense perennial vegetation, e.g. a natural forest, builds up a soil ecosystem rich in biodiversity. When an area is deforested and converted into cultivated lands, processes leading to soil erosion are triggered. Mismanagement of the soil in an area further escalates the processes that on a long-term basis end up in an unproductive desert. Vegetation cover on the soil surface, especially the natural forests with trees as the dominant members of the forest community, will be a major determinant of biodiversity conservation. Agriculture development based on sustainability principles – that is, living soil, biodiversity, and cyclic nutrient flows – will prove to be most promising for the natural biodiversity to prevail within and outside the pedosphere.

Exercises

A. Multiple Choice Questions

1. The outermost layer of the Earth composed of soil and subject to soil formation processes is known as
   a. Pedosphere
   b. Lithosphere
   c. Thermosphere
   d. Exosphere
2. What does the abbreviation SOM used in pedology stand for?
   a. Save Our Motherland
   b. Soil Organic Matter
   c. Systematic Organic Management
   d. Soil Organic Management
3. What is stable organic matter in the soil?
   a. Organic matter from dead animals
   b. Organic matter from plants
   c. Humus
   d. Decomposing organic matter
4. What soil has an appropriate proportion of the three major soil types – sand (40%), silt (40%), and clay (20%) – and has a high proportion of carbon?
   a. Sandy
   b. Clay
   c. Peaty
   d. Loam
5. Which of the following types of soil is regarded as the most ideal for crop cultivation?
   a. Loam
   b. Peaty
   c. Clay
   d. Chalky

6. Which of the following types of the soil has widest spaces between its constituent particles largest in size among all soil types?
   a. Loam
   b. Clay
   c. Sandy
   d. Peaty
7. It is black and brown, has a very high proportion of carbon, and gives soft and spongy touch. What soil type it is?
   a. Chalky
   b. Peaty
   c. Loam
   d. Clay
8. A single colony of this organism may cover approximately 9.0 square kilometers of the soil. What organism is it?
   a. Honey fungus (*Armillaria ostoyae*)
   b. Mushroom (*Agaricus bisporus*)
   c. Chytrid fungus (*Batrachochytrium dendrobatidis*)
   d. Ghost fungus (*Omphalotus nidiformis*)
9. Cyanobacteria play role in
   a. Photosynthesis
   b. Nitrogen fixation
   c. Photosynthesis and nitrogen fixation
   d. None of the above
10. Which of the following fungal species is responsible for causing wilt in *Cajanus cajan*?
   a. *Fusarium udum*
   b. *Aspergellus flavus*
   c. *Trichoderma viride*
   d. *Penicillium* sp.
11. What is true about saprophytic fungi?
   a. These are parasitic.
   b. These derive carbon and energy from the organic matter they perpetuate on.
   c. These cause root rot in tomatoes.
   d. None of the above.
12. Oxylophytes are the plants that are restricted to
   a. Alkaline soils
   b. Acid soil
   c. Neutral soils
   d. Chalky soils

13. Plants occurring in saline soils are designated as

    a. Oxylophytes   b. Lithophytes

    c. Halophytes   d. Psammophytes

14. Lithophytes grow on

    a. Sandy soils   b. Saline soils

    c. Rock surface   d. Loam soils

15. Plants that grow on rock crevices or anywhere on the rocks where detritus gets accumulated are classified as

    a. Lithophytes   b. Halophytes

    c. Psammophytes   d. Chasmophytes

16. Which of the following belongs to soil microfauna?

    a. Bacteria   b. Fungi

    c. Cyanobacteria   d. Amoeba

17. The animals adapted to digging and living underground are known as

    a. Fossorial   b. Digging animals

    c. Soiled animals   d. Amphibians

18. Soils rich in lime are more especially appropriate for the

    a. Squirrels   b. Land snails

    c. Frogs   d. Kangaroo-rats

19. *Breviceps* are

    a. Fossorial mammals   b. Soil reptiles

    c. Rain frogs   d. Fossorial insects

20. According to FAO, how much proportion of the planet's biodiversity is hosted by the soils?

    a. 85%   b. 75%

    c. 50%   d. 25%

Answers

1. **a**, 2. **b**, 3. **c**, 4. **d**, 5. **a**, 6. **c**, 7. **b**, 8. **a**, 9. **c**, 10. **a**, 11. **b**, 12. **b**, 13. **c**, 14. **c**, 15. **d**, 16. **d**, 17. **a**, 18. **b**, 19. **c**, 20. **d**

B. Short-Answer Questions

1. What do you mean by pedosphere?
2. How many types of soils are found in the world?
3. Write salient features of sandy soil?
4. What are main characteristics of peaty soils?
5. Which type of soil is considered a "perfect" and ideal soil for crop cultivation?
6. Which organisms comprise soil microflora?
7. Mention examples of soil microfauna.
8. What is the fossorial "lifestyle" some soil organisms are adapted to?
9. What roles do the earthworms play in maintaining soil fertility?
10. Name the major factors responsible for soil biodiversity erosion.

C. Long-Answer Questions

1. How is the biodiversity of "invisible world" distinguishable from that of the "visible world"? Discuss various soil biodiversity attributes vital for the nature's biodiversity.
2. Discuss soil microflora and its various functions.
3. Explain soil fauna giving various examples.
4. Elaborate on various measures for the conservation of soil biodiversity.
5. Write short notes on:
   a) Pedodiversity
   b) Functional Attributes of Soil Biodiversity
   c) Land Degradation and Soil Biodiversity Loss

References

FAO. 2015. Soils and Biodiversity: *Soils Host a Quarter of Our Planet's Biodiversity – 2015 International Year of Soils.* Rome: FAO.

Gongalsky, K. B. 2021. Soil macrofauna: Study problems and perspectives. *Soil Biology and Biochemistry*, 159: 108281. doi: 10.1016/j.soilbio.2021.108281.

IPCC, 2019: Summary for Policymakers. In: P.R. Shukla, J. Skea, E. Calvo Buendia, V. Masson-Delmotte, H.- O. Pörtner, D. C. Roberts, P. Zhai, R. Slade, S. Connors, R. van Diemen, M. Ferrat, E. Haughey, S. Luz, S. Neogi, M. Pathak, J. Petzold, J. Portugal Pereira, P. Vyas, E. Huntley, K. Kissick, M. Belkacemi, J. Malley, (eds.). *Climate Change and*

Land: An IPCC Special Report on Climate Change, Desertification, Land Degradation, Sustainable Land Management, Food Security, and Greenhouse Gas Fluxes in Terrestrial Ecosystems. IPCC.

Menta, C. 2012, Soil fauna diversity – function, soil degradation, biological indices, soil restoration. In Lameed, G. A. (ed.) *Biodiversity Conservation and Utilization in a Diverse World*. London: IntechOpen. doi: 10.5772/51091.

Narayanan, S. P., Anuja, R., Thomas, A. P. and Paliwal, R. 2022. A new species of Moniligaster Perrier, 1872 (Annelida, Moniligastridae) from India, with status revision of M. deshayesi minor Michaelsen, 1913. *Opuscula Zoologica*, 53(1):31-50. doi: 10.18348/opzool.2022.1.31.

Negi, V., Nautiyal, N. and Singh, Vir. 2018. Soil Ecology: The Basis of Sustainable Agriculture and Climate Change Mitigation. New Delhi: Educationist Press. 145 pp. ISBN:9789386283771.

Sample, Ian. 2007. Global food crisis looms as climate change and population growth strip fertile land. *The Guardian*, August 31, 2007.

Singh, V. 2019. *Fertilizing the Universe: A New Chapter of Unfolding Evolution*. London: Cambridge Scholars Publishing. 285pp. (ISBN (10): 1-5275-2301-2 ISBN (13): 978-1-5275-2301-2).

Web Resources

https://naldc.nal.usda.gov/download/IND43893638/PDF

https://ourworldindata.org/land-use

https://www.intechopen.com/chapters/38612

https://www.ipcc.ch/site/assets/uploads/2019/11/02_Summary-for-Policymakers_SPM.pdf

https://www.shareyouressays.com/essays/soils-fauna-and-soil-flora-essay/121329

https://www.theguardian.com/environment/2007/aug/31/climatechange.food

https://www.unccd.int/sites/default/files/relevant-links/2017-01/UNCCD_Convention_ENG_0.pdf

https://www.undp.org/content/undp/en/home/sustainable-development-goals/goal-15-life-on-land/targets.html

https://www.universetoday.com/25756/surface-area-of-the-earth/

4

Biodiversity in the Chemosynthesis-Based Community

Nutrition is the basis for life to prosper and sustain in the biosphere. There are two modes of nutrition in all species: self-nourishment and cross-nourishment. Self-nourishment goes on in the category of organisms known as autotrophs. Cross-nutrition is maintained by the category of consumers categorized as heterotrophs. The autotrophs nourish themselves by utilizing inorganic molecules and energy available in nature, while the heterotrophs derive the nutrients that the autotrophs 'manufacture' for themselves. In other words, the heterotrophs derive nutrients from organic molecules. These organic molecules are derived either from the autotrophs directly or from the consumers dependent on autotrophs.

Autotrophs derive energy in its free form (e.g. light) or that available in inorganic molecules (e.g., CO, H_2S, Fe, NH_4^+, etc.), while the heterotrophs harness energy bound with organic molecules. In essence, the process of nourishment the heterotrophs apply for their survival, growth, and sustenance involves the energy of organic molecules derived from the autotrophs.

The Two Types of the Living Communities

All living organisms, based on their nutrition, can be categorized into two types of communities:

i. Photosynthesis-based communities, and

ii. Chemosynthesis-based communities

Communities based on photosynthesis involve all the terrestrial communities, most of the organisms in the soil and aquatic ecosystems, whereas communities dependent on chemosynthesis are there in the soils, cold seeps, sunken woods, vertebrate falls (e.g., whale carcasses), and on the ocean floor around hydrothermal vents.

The Photosynthesis Based Communities

Photosynthesizing organisms get nourished using inorganic substances CO_2 and H_2O synthesized together into sugars in the presence of light and chlorophyll. What is most crucial in the synthesis of carbohydrates is the role of energy in the form of light. Light, in essence, is the energy that is converted into biochemical energy, or energy of life, or living energy. The living energy or the biochemical energy eventually sourced into the light is bonded with sugars in living organisms. Photosynthesis, putting otherwise, is a process in which light transforms itself into energy binding together CO_2 and H_2O within chloroplasts found in the green plants, algae, and cyanobacteria, and releasing oxygen molecules. Thus, only chlorophyllous organisms are capable of performing photosynthesis and staying as autotrophs.

Solar energy trapped by chlorophyllous organisms in the process of photosynthesis is utilized for the synthesis of all biomolecules, for their maintenance, growth, and reproduction. In addition, this energy also supports a huge number of heterotrophic organisms in communities ranging from herbivores to top carnivores. The energy flow from autotrophs – the producers – to subsequent trophic levels constituted by consumers (herbivores, primary carnivores, secondary carnivores, and top carnivores), following the 10% Law enunciated by Lindeman (1942), becomes the basis for nourishing an enormous number and varieties of organisms. A significant proportion of the organic matter produced by living photosynthesizers and from dead organisms (producers as well as consumers) forms detritus that also supports a huge number and numerous varieties of detritus consumers in the soils and the sediment in aquatic ecosystems (Fig. 4.1).

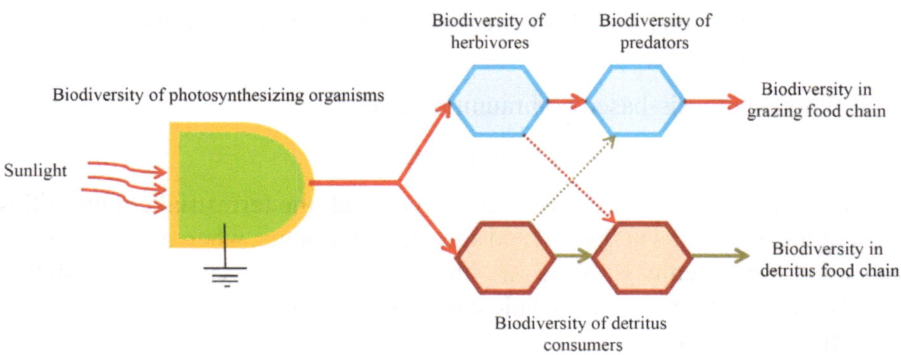

Fig. 4.1: Photosynthesis fuels biodiversity in grazing as well as in detritus food chains

Photosynthesis, thus, feeds most of the life on the planet in terrestrial, soil, and aquatic ecosystems. Among communities of the Earth's ecosystems, in this way, photosynthesis spells out enormous biodiversity at Alpha, Beta, and Gamma levels (Fig. 4.2). Each trophic level and each food chain – grazing as well as detritus – in the photosynthesis-dependent communities in all ecosystems represent a very high measure of biodiversity. The photosynthesis-based biodiversity undoubtedly comprises a greater proportion of the planet's life. The humanity is exclusively dependent on photosynthesis for survival and sustenance.

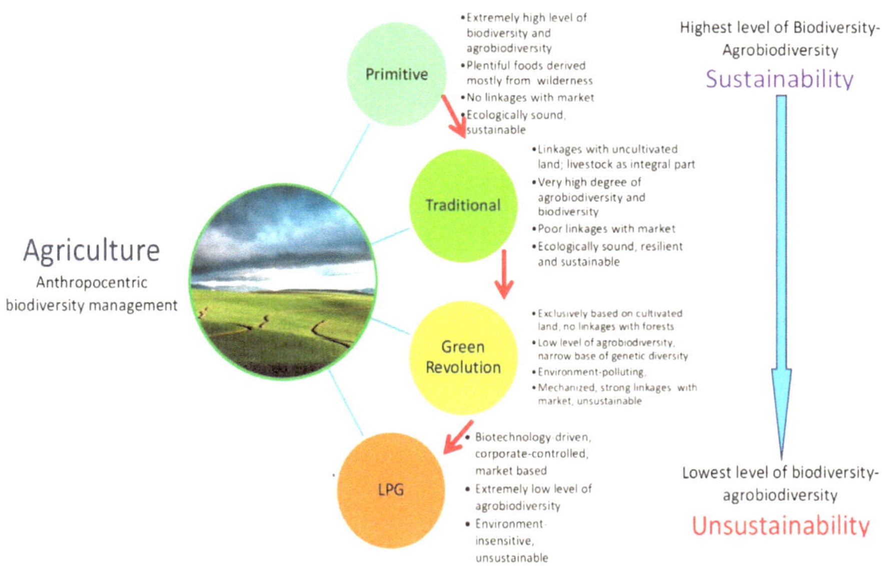

Fig. 4.2: Photosynthesis-dependent biodiversity blossoms in terrestrial ecosystems, photic zones of aquatic ecosystems, and soil ecosystems

The Chemosynthesis Based Community

The other category of the community comprises the organisms dependent on chemosynthesis. These are specialized microorganisms, especially bacteria and archaea. Chemosynthesis-based communities were discovered in the late 1970s. In the field of biology, this discovery was intriguing. There are three major locations/ ecosystems the chemosynthesis-based communities have evolved in and/ or may develop in: hydrothermal vents, cold seeps, and hot springs on land. Whale fall on the ocean floor also becomes a habitat for the chemosynthetically fuelled community. Sunken wood and soil communities are also partially chemosynthesis-dependent. The earliest-known chemosynthetic association containing metazoan animals has been described from the Silurian of Ural Mountains (Kaim 2010).

Chemosynthetic organisms synthesize complex organic compounds utilizing inorganic substances as a source of energy and CO_2 or CH_4 as carbon sources. Such autotrophs are also referred to as chemoautotrophs. Chemosynthesis occurs in the absence of light. chemoautotrophs inhabit places/ ecosystems away from the reach of solar radiation. Fixing carbon by using the oxidation of chemical compounds, the chemoautotrophs are phylogenetically diverse, representing taxa of biochemical interest, e.g., gammaproteobacteria, epsilon proteobacteria, aquificaeles, neutrophilic iron-oxidizing bacteria and methanogenic archaea (Singh 2020).

The term – chemosynthesis was coined in 1897 by Wilhelm Pfeffer. Before Wilhelm Pfeffer, Sergei Nikolaevich Vinogradskii had revealed that some microorganisms were capable of living exclusively on inorganic substances. He referred to this process of life as “anorgoxydant” in 1890, which later on was known as chemosynthesis. This auto nourishment by specialized microorganisms is also known as chemoautotrophy. More recently, this process has been assigned the term chemolithoautotrophy.

A variety of chemosynthetic organisms prevails in nature depending upon the chemical substances they oxide to use as energy sources. The inorganic substances the chemosynthetic organisms derive energy from are: H_2, CO, Fe^{2+}, NH_4^+, NO_2^-, H_2S, S, etc. Nearly entire biological nitrogen fixation occurs due to the role of certain chemosynthetic microbes thriving independently in soils, waters, and synergistically in the roots of many plants, especially belonging to the Fabaceae or the Leguminosae family. The nitrifying bacteria are of critical ecological importance as much of the ecosystem structure is owing to the specific role of these bacteria in nitrogen fixation. Nitrogen incorporated into amino acids and then into protein is a basic structural molecule. These microorganisms oxidize NH_4^+ to NO_2^- and subsequently to NO_3^- and derive energy for the synthesis of complex biomolecules:

$$2NH_4^+ + 3O_2 \rightarrow 2NO_2^- + 4H^+ + 2H_2O + \text{Energy}$$

Cold seep areas widespread around the globe’s oceans are characterized by the presence of fluxes of methane, oil, and other hydrocarbons that make their way to the sediment-water interface from below via relatively focused conduits, such as faults (Arvidson et al. 2004). These areas are “hot spots” for chemosynthesis with methane and its decomposition by-products like hydrogen sulfide (Milucka et al. 2012). The chemosynthesis profusely going on in the darkest ocean zone at about 2 °C – 4 °C gives rise to, supports, and sustains biotic communities (Das et al. 2011, Giongo et al. 2015). Northern Atlantic and Pacific Oceans have been extensively covered in the studies focused on chemosynthetic communities (Barry et al. 2000, Genio et al. 2008, Das et al. 2011, Vigneron et al. 2013a,b, Giongo et al. 2015). Fewer studies have also been conducted in western Africa and Peru, Chile, and New Zealand (Rodrigues et al. 2010, Giongo et al. 2015). However, there was very scant information about the chemosynthetic communities prevailing in the western South Atlantic Sea until Giongo et al. (2015) conducted extensive studies in this region.

Some of the striking features of the chemosynthetic kind of communities that have been recorded and documented by Giongo et al. (2015) are: mat-forming vacuolated sulfur-oxidizing bacteria *Beggiatoa* (a genus of Gammaproteobacteria), *Arcobacter* (a genus of Gram-negative, spiral-shaped bacteria), *Thiothrix* (a genus of filamentous sulfur-oxidizing bacteria), and *Thiomargarita* (a genus of family Thiotrichaceae which includes vacuolated sulfur bacteria species). The common animal species existing amid chemosynthetic communities include polychaete annelids belonging to the 'Vestimentifera' clade belonging to the family Siboglinidae. Interactive relationships among various members of a chemosynthetic community involve direct predation as well as endosymbiosis. The ecological roles of the free-living chemosynthetic organisms and symbionts are pretty interesting but still need to be understood in their entirety.

Biodiversity in Hydrothermal Vent Ecosystems

Chemosynthetic communities profusely prevail on ocean floors around the globe. Before the discovery of the unique living systems flowering around hydrothermal vents on the ocean floor in the Galápagos Rift in the Pacific Ocean in 1977, it was assumed that photosynthesis is the sole source of energy for ocean life.

This discovery of a new kind of life flourishing on the floor of the Pacific Ocean amidst total darkness, albeit accidental, nonplussed the ecologists. This non-targeted but fascinating discovery led to unfold a mystery of life.

Hydrothermal vents might be active, inactive, and somewhere between. Active vent zones, relatively less abundant on seafloors, generally occur in areas of underwater volcanic activity. Fluids boiling at >350 °C ooze out of the Earth's crust via cracks in the seabed, ejecting into the water column through smoking chimneys. There is immense biodiversity nourished through chemosynthesis in and around these active smokers.

The chemosynthesis-based life in the complete darkness of the ocean bottom persists in the warm water having sulfur compounds discharged from the hydrothermal vents. The elemental sulfur and sulfur compounds serve as energy sources for the bacteria to carry out chemosynthesis. Elemental sulfur, hydrogen sulfide, or thiosulfate present in the hydrothermal vents' discharges undergo oxidation, and the energy released is efficiently utilized by the sulfur-oxidizing bacteria in the processes of carbohydrate synthesis fixing carbon in CO_2 dissolved in water (Singh 2020):

$$2H_2S + O_2 \rightarrow 2S^o + 2H_2O + \text{Energy}$$

$$6CO_2 + 6H_2O + 3H_2S + \text{Energy} \rightarrow C_6H_{12}O_6 + 3H_2SO_4$$

There is only one chemical reaction that represents photosynthesis. The chemosynthesis, however, may be represented by several chemical reactions based on the inorganic substance as an energy source.

The sulfur-oxidizing bacteria in the hydrothermal biotic community are free-living and those dwelling symbiotically in the tissues of many invertebrates. Giant tubeworm (*Riftia pachyptila*) (Fig. 4.3) is one of the well-studied invertebrate species within the chemosynthetic hydrothermal community. A giant tubeworm could be as longer than 150 cm and 4 cm in diameter. No mouth, no eyes, no legs, no alimentary tract—this chemosynthesis-dependent animal is a baffling creature in itself.

The giant tubeworm belongs to the family of polychaete annelid worms. This marine invertebrate has green-brown spongy tissue known as trophosome containing specialized cells that remain packed with chemosynthetic microbes – the sulfur-oxidizing bacteria – numbering about 10 billion per gram of the trophosome and contributing 60% of the total biomass of the host. The upper portion of the invertebrate, the plume, attains red color owing to the presence of hemoglobin. With hydrogen sulfide available plentifully in the hydrothermal vents and oxygen dissolved in water, the plume generates energy to utilize carbon from the CO_2 in the marine environment for synthesizing biomolecules. Chemosynthesis occurring within the trophosome was recognized by American microbiologist Colleen Cavanaugh and her colleagues (Cavanaugh et al. 1981). It is with this mechanism that the tubeworms are nourished following a symbiotic relationship with the chemosynthetic organisms. The outer covering of the giant tubeworm, which is made up of chitin, helps the animal survive extreme pressure and boiling-to-freezing range of temperatures.

Fig. 4.3: The giant tubeworms (*Riftia pachyptila*) from the 2011 NOAA Galápagos Rift Expedition. (Source: The image is in the public domain as mentioned on https://commons.wikimedia.org/wiki/File:Riftia_tube_worms_Galapagos_2011.jpg)

The tubeworm in the hydrothermal ecosystem functions as a keystone species playing a critical role in maintaining the structure of this chemosynthesis-fuelled community. The tube of a worm protects it from predators. However, several deep-sea crustaceans feed on the plume part of the worms. The predators prospering around tubeworms, thus, also belong to the chemosynthetic society. The chemosynthetic communities dependent on sulfur-oxidizing bacteria also prosper around the thermal vents in deep freshwater lakes, caves, as well as in surface hot springs. A hydrothermal vent community involving producers (free-living chemosynthesizers and those associated with the tubeworm), and various trophic levels of carnivores can be represented through an ecological pyramid – akin to an energy pyramid (Fig. 4.4).

Since the discovery of the deep-sea hydrothermal vent life in the Galápagos Islands in 1977, hundreds of hydrothermal vents and over 500 animal species surviving around hydrothermal vents have been discovered. The major groups of species include Eubacteria (Strain NS-E), Archaebacteria (Pyrococcus strain GB-D and *Thermococcus fumicolans*), Protists (Apicomplexa, Perkinsozoa, Syndiniales, and Kinetoplastida), plant (dandelions), and animals (Tubeworms, Clams, Mussels, Crabs, Shrimp, Zoracid fish, Octopus, etc.). The submarine hydrothermal vents are biologically more diverse and productive. The thriving biotic communities represent a very high degree of species endemism.

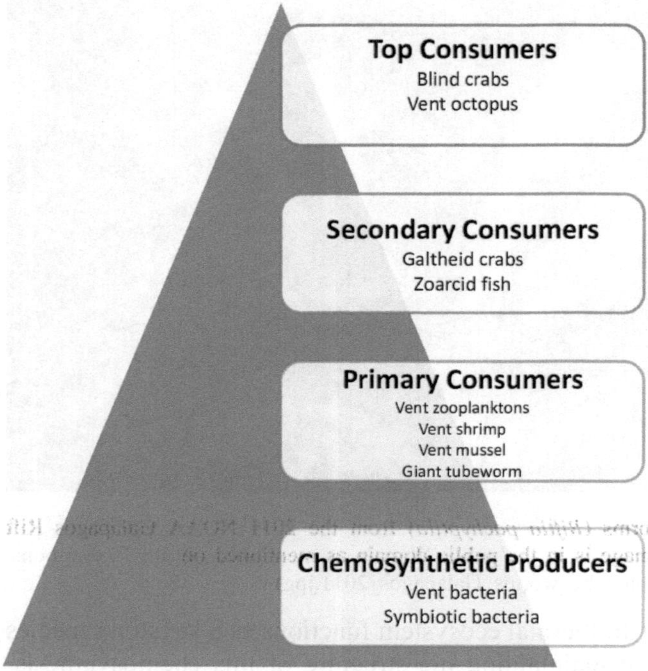

Fig. 4.4: An ecological pyramid representing a deep-ocean chemosynthesis-dependent hydrothermal community

Threats to the Chemosynthesis-Based Biodiversity

Although largely invisible and not easily accessible, the biodiversity bloom attributable to chemosynthesis is facing crises caused and exacerbated by anthropogenic activities. The deep-sea hydrothermal vent ecosystems are exceptionally rich in mineral deposits. The minerals, some of them of exceptionally high commercial value, are to be extensively exploited by some companies exclusively for private profits. Although in its infancy, the deep-sea mining industry has tremendous potential to irreversibly destroy the hydrothermal vent ecosystems and rich biodiversity thriving therein. Many private companies with expertise in deep-sea mining have already won contracts for exploiting valuable wealth deposited within the living communities.

Mining activities on and near an active vent, which measure only about 50 km^2 in total, would lead to the destruction of living organisms. Losses of biodiversity would be severe and irreversible due to alteration in the vulnerable habitats.

Another risk the chemosynthesis-dependent biodiversity faces is that several species endemic to hydrothermal vent ecosystems are attracting the attention of the pharmaceutical industry. Some companies appear to be in the stage of planning species exploitation for manufacturing medicines and other products of extraordinary socioeconomic value. Once large-scale exploitation of the mineral resources and living species begins, the ecosystems blossoming like oases on the ocean floor would be in perilous conditions.

Till now, it was readily believable that the deep-sea creatures would be safe from the ongoing Sixth Mass Extinction against the backdrop of Anthropocene. But, it is now well understood that even the seemingly invisible species are not away from the grips of the human-led extinction processes.

International Obligations

Hydrothermal vent zones are found lying on the international seabed beyond national jurisdiction. The UN Law of the Sea Treaty has established an organization known as International Seabed Authority (ISA) to look after the matter relating to and various issues associated with hydrothermal vents. The ISA formulates regulations governing the mining of the vent zones.

The rarity and vulnerability of hydrothermal vent life have received global attention and recognition only in recent years. Parties of the Convention on Biodiversity (COB) held in Rio de Janeiro in 1992 gave recognition to hydrothermal vents as "ecologically and biologically significant areas" in need of enhanced conservation and management measures. The United Nations General assembly has given a call for States to manage risks to the marine biodiversity of hydrothermal vents.

As the chemosynthesis-fed biodiversity is the last remaining component of relatively undisturbed biodiversity away from human's direct accessibility, this needs to be given due protection. Hydrothermal vents' biodiversity is "virgin biodiversity" which needs to be left only as "virgin". The history of the Evolution of life is hidden in this "virgin biodiversity"—which is yet to unfold. Once disturbed, damaged, and/ or destroyed, the biodiversity blossoming at its ecological climax amidst deep-sea active hydrothermal vents will perish. And with this gloomy state of the "virgin life", all the mysteries of Evolution will also vanish forever.

Protection and conservation of the chemosynthesis-fuelled biodiversity cannot be ensured only by formulating and promulgating international regulations and laws. All the concerned nations must constitute their own rules and regulations that could ensure that this pristine biodiversity keeps flowering on ocean floors as it has been for millions of years as per nature's Evolution's designs. This biodiversity and the ecosystems infusing life in it should be kept away from exploitation for economic purposes. It is not only necessary but also imperative in our own life. It is the supreme obligation on the part of humanity.

Summary

All living organisms, based on their nutrition, can be categorized into two types of communities: photosynthesis-based, and chemosynthesis-based. Communities based on photosynthesis involve all the terrestrial communities, most of the organisms in the soil and aquatic ecosystems, whereas communities dependent on chemosynthesis are there in the deep-sea waters, vertebrate falls (e.g., whale carcasses), and on the ocean floor around hydrothermal vents, and partially in the soils and sunken woods. Photosynthesis generates, supports, and sustains most of the biodiversity on the planet including entire humanity as an integral part of it. Chemosynthetic organisms referred to as chemoautotrophs or chemolithoautotrophs, synthesize complex organic compounds utilizing inorganic substances as various sources of energy and CO_2, or CH_4 as carbon sources. The chemoautotrophs are phylogenetically diverse, representing taxa of biochemical interest, e.g., gammaproteobacteria, epsilon proteobacteria, aquificales, neutrophilic iron-oxidizing bacteria, and methanogenic archaea.

A variety of chemosynthetic organisms prevails in nature depending upon the chemical substances they oxide to use as energy sources. The inorganic substances the chemosynthetic organisms depend on are: H_2, CO, Fe^{2+}, NH_4^+, NO_2^-, H_2S, S, etc. Nitrogen-fixing microorganisms oxidize NH_4^+ to NO_2^- and subsequently to NO_3^- and derive energy for the synthesis of complex biomolecules:

$$2NH_4^+ + 3O_2 \rightarrow 2NO_2^- + 4H^+ + 2H_2O + \text{Energy}$$

Cold seep areas widespread around the globe's oceans are characterized by the presence of fluxes of methane, oil, and other hydrocarbons that make their way to the sediment-water interface from below via relatively focused conduits, such as faults. The chemosynthesis profusely going on in the darkest ocean zone at about 2 °C–4 °C gives rise to, supports, and sustains biotic communities. Some of the striking features of the chemosynthetic kind of communities are: mat-forming vacuolated sulfur-oxidizing bacteria *Beggiatoa* (a genus of Gammaproteobacteria), *Arcobacter* (a genus of Gram-negative, spiral-shaped bacteria), *Thiothrix* (a genus of filamentous sulfur-oxidizing bacteria) and *Thiomargarita* (a genus of family Thiotrichaceae which includes vacuolated sulfur bacteria species). The common animal species existing amid chemosynthetic communities include polychaete annelids belonging to the 'Vestimentifera' clade that belongs to the family Siboglinidae.

The chemosynthesis-based life in the complete darkness of the ocean bottom persists in the warm water having sulfur compounds discharged from the hydrothermal vents. The elemental sulfur and sulfur compounds serve as energy sources for the bacteria to carry out chemosynthesis. Elemental sulfur, hydrogen sulphide, or thiosulfate present in the hydrothermal vents' discharges undergo oxidation, and the energy efficiently utilized by the sulfur-oxidizing bacteria in the processes of carbohydrate synthesis fixing carbon in CO_2 dissolved in water:

$$2H_2S + O_2 \rightarrow 2S^o + 2H_2O + \text{Energy}$$

$$6CO_2 + 6H_2O + 3H_2S + \text{Energy} \rightarrow C_6H_{12}O_6 + 3H_2SO_4$$

The sulfur-oxidizing bacteria in the hydrothermal biotic community are free-living and symbiotically dwell in the tissues of many invertebrates. Giant tubeworm (*Riftia pachyptila*) is one of the well-studied invertebrate species within the chemosynthetic hydrothermal community. The giant tubeworm belongs to the family of polychaete annelid worms. The sulfur-oxidizing bacteria – numbering about 10 billion per gram of the trophosome and contributing 60% of the total biomass of the host – function as the exclusive energy source for the invertebrate. The upper portion of the invertebrate, the plume, attains red color owing to the presence of hemoglobin. With hydrogen sulfide available plentifully in the hydrothermal vents and oxygen dissolved in water, the plume generates energy to utilize carbon from the CO_2 in the marine environment for synthesizing biomolecules. Hundreds of hydrothermal vents and over 500 animal species prospering around hydrothermal vents have been discovered. The major groups of species include Eubacteria (Strain NS-E), Archaebacteria (Pyrococcus strain GB-D and *Thermococcus fumicolans*), Protists (Apicomplexa, Perkinsozoa, Syndiniales, and Kinetoplastida), plant (dandelions), and animals (Tubeworms, Clams, Mussels, Crabs, Shrimp, Zoracid fish, Octopus, etc.).

Although the chemosynthetically fuelled biodiversity is prospering in the ecosystems devoid of light and largely invisible to human beings, this wonderful diversity is facing crises due to human activities, particularly relating to deep sea mining. The chemosynthesis-dependent communities need to be protected and conserved following international and national laws and by evolving biodiversity conservation strategies appropriate for chemosynthesis-based communities.

Exercises

A. Multiple Choice Questions

1. How many communities based on the source of energy their organisms utilize prevail in the world?

   a. 2 b. 4
   c. 8 d. 12

2. What two material sources are required for photosynthesis?

   a. CO_2 and H_2O b. CO_2 and O_2
   c. Light and chlorophyll d. Light and ATP

3. Which of the following are photoautotrophs?

   a. Mushrooms
   b. Green plants, algae, and cyanobacteria
   c. Cuscuta plant
   d. Paramecium

4. Which of the following photosynthesizers also performs nitrogen fixation?

   a. A mango tree b. Blue-green algae
   c. Pitcher plant d. Pigeon pea

5. Who enunciated the 10% Law of energy transfer?

   a. Odum b. Ernst Haeckel
   c. Lindeman d. Hutchinson

6. Photosynthesis-dependent biodiversity prevails in

   a. Terrestrial communities
   b. Aquatic communities
   c. Terrestrial as well as aquatic communities
   d. Hydrothermal vent communities

7. Energy source in a detritus food chain emanates from

   a. Autolithochemotrophy b. Autophototrophy
   c. Both autolithotrophy and photoautotrophy d. Photosynthesis

8. Which of the following statements is right?
   a. Chemosynthesis occurs in the absence of light.
   b. Whale carcasses support chemosynthetic organisms.
   c. Photosynthesis is represented by a single equation, chemosynthesis by many.
   d. All of the above.
9. Cold seep areas harbor biodiversity based on
   a. Chemosynthesis
   b. Photosynthesis
   c. Both chemosynthesis and photosynthesis
   d. None of the above
10. Which of the following parts of a giant tubeworm contains chemosynthetic bacteria?
    a. Plume
    b. Trophosome
    c. Chitin cover
    d. None of the above
11. What chemosynthetic bacteria are the following equations related with?

    $2H_2S + O_2 \rightarrow 2S^0 + 2H_2O + Energy$

    $6CO_2 + 6H_2O + 3H_2S + Energy \rightarrow C_6H_{12}O_6 + 3H_2SO_4$

    a. Fe-oxidizing bacteria
    b. Mn-oxidizing bacteria
    c. Sulfur-oxidizing bacteria
    d. All of the above
12. What part of the giant tubeworm contains hemoglobin?
    a. Trophosome
    b. Chitin cover
    c. Both trophosome and chitin cover
    d. Plume
13. Which organisms are the top consumers in a hydrothermal vent ecosystem?
    a. Vent shrimps and mussels
    b. Giant tubeworms
    c. Vent zooplanktons
    d. Blind crabs and vent octopus

14. Life in a hydrothermal vent on the ocean floor was first of all encountered in 1977 in
    a. The Andaman Sea of the Indian Ocean
    b. Bay of Bengal
    c. Galápagos Rift in the Pacific Ocean
    d. Mediterranean Sea
15. How much proportion of the biomass of a giant tubeworm is contributed by symbiotic chemosynthetic bacteria?
    a. 35% b. 60%
    c. 80% d. 90%
16. At what temperature do the fluids discharge from the Earth's crust via cracks in the seabed, ejecting into the water column through smoking chimneys?
    a. 350 °C b. 150 °C
    c. 100 °C d. 37 °C
17. What poses the most serious threat to the chemosynthesis-dependent biodiversity in deep-sea waters?
    a. Deep-sea mining industry b. Scientific research
    c. Naval operations d. Exotic carnivorous species
18. *Riftia pachyptila* is the zoological name of the
    a. Vent mussels b. Giant tubeworms
    c. Blind crabs d. Vent octopus
19. The UN Law of the Sea Treaty has established an organization known as
    a. International Seabed Agency
    b. International Seabed Law
    c. International Seabed Authority
    d. International Ocean Biodiversity Conservation Law
20. Which of the following is a true statement?
    a. Parties of the Convention on Biodiversity (COB) held in Rio de Janeiro in 1992 gave recognition to hydrothermal vents as "ecologically and biologically significant areas" in need of enhanced conservation and management measures.

b. The ISA formulates regulations governing the mining of the vent zones.

c. Hundreds of hydrothermal vents and over 500 animal species prospering around hydrothermal vents have been discovered.

d. All of the above

Answers: 1. **a**, 2. **a**, 3. **b**, 4. **b**, 5. **c**, 6. **c**, 7. **d**, 8. **d**, 9. **a**, 10. **b**, 11. **c**, 12. **d**, 13. **d**, 14. **c**, 15. **b**, 16. **a**, 17. **a**, 18. **b**, 19. **c**, 20. **d**

B. Short-Answer Questions

1. What are the sources of energy for all living organisms in the biosphere?
2. What are the differences between a photosynthesis-based and a chemosynthesis-based community?
3. Mention some examples of the chemosynthesis-dependent communities.
4. What are the sources that the chemosynthetic organisms use for releasing energy for carbohydrate synthesis?
5. Name some of the animals that prosper in a chemosynthetic community in deep-sea waters.
6. What is found in the fluids that ooze out of a hydrothermal vent?
7. When and where was the life in a hydrothermal vent encountered?
8. Which part of a giant tubeworm contains hemoglobin?
9. In which part of a tubeworm do the symbiotic bacteria occur and perform chemosynthesis?
10. Name the organization formulated by UNO that has an obligation for regulating deep-sea mining.

C. Long-Answer Questions

1. Discuss the energy sources based on the basis of all living organisms on Earth can be categorized into communities. How does biodiversity thrive in these communities fulfilling their energy needs?
2. Explain the major features of the chemosynthesis-based communities in deep-sea waters.
3. Discuss with the help of an ecological pyramid the structure and functioning of a hydrothermal vent community.
4. How can the chemosynthesis-fuelled biodiversity in the oceans be protected and conserved? Give your suggestions.

5. Write short notes on the following:
    a) The Giant Tubeworm (*Riftia pachyptila*)
    b) Photosynthesis-Dependent Biodiversity
    c) Threats to Hydrothermal Vent Biodiversity

References

Arvidson, R. S., Morse, J. W. and Joye, S. B. 2004. The sulfur biogeochemistry of chemosynthetic cold seep communities, Gulf of Mexico, USA. *Mar. Chem.*, 87: 97-119.

Barry, J. P., Buck, K. R., Goffredi, S. K. and Hashimoto, J. 2000. Ultrastructure studies of two chemosynthetic invertebrate-bacterial symbiosis (*Lamellibrachia* sp. and *Acharax* sp.) from Hatsushima cold seeps in Sagami Bay, Japan. *Deep-Sea Res.* PT II, 16: 91-99.

Cavanaugh, C. M., Gardiner, S. L., Jones, M. L., Jonnasch, H. W. and Waterbury, J. B. 1981. Prokaryotic cells in the hydrothermal vent tube worm Riftia pachyptila: possible chemoautotrophic symbionts. Science, 213 (4505): 340-342. doi: 10.1126/ science.213.4505.340.

Das, A., Sujith, P. P., Mourya, B. S., Biche, S. U. and LokaBharathi, P. A. 2011. Chemosynthetic activity prevails in deep-sea sediments of the Central India Basin. *Extremophiles*, 15: 177-189.

Genio, L., Johnson, S. B., Vrijenhoek, R. C., Cunha, M. R., Tyler, P., et al. 2008. New record of *Bathymodiolus mauritanicus* Cosel 2002 from Gulf of Cadiz (NE Atlantic) mud volcanoes. *J. Shellfish Res.*, 27: 53-61.

Giongo, A., Haag, T., Simão, T. L. L., Medina-Silva, R., Utz, L. R. P., Bogo, M. R., Bonatto, S. L., Zamberlan, P. M., Augustin, A. H., Lourega, R. V., Rodrigues, L. F., Sbrissa, G. F., Kowsmann, R. O., Freire, A. F. M., Miller, D. J., Viana, L. F., Ketzer, J. M. M. and Eizirik, E. 2015. Discovery of a chemosynthesis-based community in the Western South Atlantic Sea. *Deep Sea Research Part I.* Link: http://dx.doi.org/10.1016/i.dsr.2015.10.010.

Kaim, A. 2010. Chemosynthesis-based communities through time. *Geophysical Research Abstracts*, 12 EGU2010-12679-1.

Lindeman, R. L. 1942. The trophic-dynamic aspect of ecology. *Ecology*, 23 (4): 399-418. doi:10.2307/1930126.

Milucka, J., ferdelman, T. G., Polerecky, L., Franzke, D., Wegener, G., Schmid, M., Lieberwirth, I., Wagner, M., Widdel, F. and Kuypers, M. M. M. 2012. Zero-valent sulphur is a key intermediate in marine methane oxidation. *Nature*, 491: 541-546.

Rodrigues, C. F., Webster, G., Cunha, M. R., Duperron, S. and Weightman, A. J. 2010. Chemosynthetic bacteria found in bivalve species from mud volcanoes of the Gulf of Cadiz. *FEMS Microbiol. Ecol.*, 73: 486-499.

Singh, V. 2020. *Environmental Plant Physiology: Botanical Strategies for a Climate Smart Planet*. CRC Press (Taylor and Francis) (ISBN: 9780367030421)

Vigneron, A., Cruaud, P., Pignet, P., Caprais, J. C., Cambon-Bonavita, M. A., Godfroy, A. and Toffin, L. 2013a. Archaeal and anaerobic methane oxidizer communities in the Sonora margin cold seeps, Guaymas Basin (Gulf of California). *ISME J.* 7: 1595-1608.

Vigneron, A., Cruaud, P., Pignet, P., Caprais, J. C., Gayet, N., Cambon-Bonavita, M. A., Godfroy, A. and Toffin, L. 2013b. Bacterial communities and syntrophic associations involved in anaerobic oxidation of methane process of the Sonora Margin cold seeps, Guaymas Basin. *Environ. Microbiol.*, 16: 2777-2790.

Web Resources

https://commons.wikimedia.org/wiki/File:Riftia_tube_worms_Galapagos_2011.jpg

https://funny-animalz.blogspot.com/2021/07/riftia-pachyptila-giant-tube-worm-size.html

https://futureoceans.earthjournalism.net/spotlight-biodiversity-and-resilience-of-deep-sea-hydrothermal-vent-communities/index.html

https://www.pewtrusts.org/-/media/assets/2019/09/deep-sea_mining_on_hydrothermal_vents_fact_sheet_v1.pdf

https://www.researchgate.net/publication/234392079_Chemosynthesis-based_communities_through_time

5

Threats to Biodiversity

A large number of genetic resources, species, and ecosystems are under severe threat. Many species have gone extinct and many more are being pushed towards extinction. Most of the ecosystems have also gone into what we can call functional extinction. The major factors becoming threats to biodiversity are anthropogenic. Human activities, in essence, are not benevolent for biodiversity to flourish in natural systems. Burgeoning human populations are the root cause of biodiversity destruction. Changing human lifestyles fed by a materialistic economy are further acting as catalysts for the ever-increasing human populations to destroy biodiversity evolved over millions of years amid terrestrial conditions favorable for the origin and prevalence of an astonishing variety of life forms on Earth.

There is hardly any threat to biodiversity due to natural factors except those inherent in the designs of natural evolution. A species at a certain point of time in its evolutionary history is destined to vanish, but the void in nature created due to this loss is filled by the evolution of new life forms. Not only the void is filled, the nature gradually goes on enriching itself with more life in tune with the pace of evolution. The biosphere has become vibrant with life thanks to the benevolence of the evolutionary journey. The evolutionary processes in our contemporary times, however, seem to be gradually hampered by anthropogenic activities and, as a result, the biodiversity in the biosphere is existing under severe stress. The major factors threatening biodiversity are described below.

Destruction of Habitats

The primary threat to biodiversity lies in habitat destruction. All living species and their diverse genotypes prosper in their specific habitats. There are now only countable ecosystems that are still safe from human intervention. There is hardly a virgin forest left on Earth. Almost all ecosystems are now 'explored' by human beings. Almost all the habitats have been modified to suit the burgeoning requirements of the human species. Forest ecosystems have been transformed into cultivated land, grazing lands, grasslands, mining sites, etc. What used to be habitats of a variety of wild animals are now human settlements – villages, towns, cities, commercial and industrial establishments, playgrounds, tourist spots, honeymoon huts, and the likes.

When habitats providing refuge to a variety of flora and fauna are destroyed and/ or transformed exclusively for human use, wild animals and endemic flora suffer in various ways. Food chains break down, animals do not get adequate food, breeding grounds of wild animals are disrupted, the breeding capacity of animals declines, routes of migratory animals are disturbed, scared animals have to survive under severe stress, and native plants do not grow and prosper following their natural vigor. Habitat destruction even of small measure breaks down symbiotic relationships among different species of plants, animals, and microorganisms. Populations of the species go on the decrease and the modified environment throws many a species into the vicious cycle of extinction.

Habitat Fragmentation

Habitat fragmentation means dividing a large intact habitat into two or multiple fragments. The fragmentation of habitats in our times is almost inevitable owing to several developmental activities that involve large areas of habitats, for example, roads, canals, railway tracts, lawns, powerlines, etc. Habitat fragmentation threats biodiversity in many ways, viz. i) by limiting dispersal and colonization potential of species; ii) by reducing the foraging ability of the wild animals; iii) by disturbing ecological niches and/ or separating species from their niches.

Introduction of Exotic Species

All the species and their genotypes in a community co-evolve in a common environment over millennia, establishing relationships of interdependence, symbiosis, and synergism. A community generally does not allow an exotic or an alien species to enter into its boundary (ecosystem) unless some adverse factors are prevailing amidst the community. Most of the exotic species also do not tend to establish in a new community. However, if a species establishes in a new environment, it would do so with disrupting effects. The introduction of an exotic biota can be intentional or accidental. In both cases, ecological effects can be deleterious. The exotic species exert pressure on a new community in several ways and may lead to serious ecological and environmental consequences. The exotic species in an ecosystem may:

i. alter ecosystem structures and species composition;

ii. induce alterations in habitats to the extent that native species would no longer be able to persist;

iii. release allelochemicals detrimental to many native species and soil microflora;

iv. prey upon native wild species to the extent of extinction;

v. cause epidemics (e.g., some microorganisms) and eliminate native species;

vi. act as causal agents of some communicable diseases giving a jolt to public health;

vii. dramatically affect functional attributes of an ecosystem, such as biodiversity conservation, photosynthetic efficiency, nitrogen fixation rates, and primary and secondary productivity.

The ecological effects of exotic species on island ecosystems – that are home to most of the world's rare and threatened species – are especially grave.

Once colonized in these ecosystems, the exotic species are destined to replace endemic species.

Some of the examples of exotic species which have caused a dramatic nuisance in ecosystems and are often talked about are *Parthenium, Lantana, Eupatorium*, *Eichhornia*, etc. *Lantana camara* shrub, a native to the West Indies, flowers throughout the year and has very high fecundity. This was introduced during the British period as an ornamental plant. This shrub has replaced native species of the sub-tropical forests in the outer Himalayas and is now spreading even in the temperate Himalayan region. Native species in large numbers have vanished due to the extensive expansion of this weed.

Parthenium hysterophorus, a native of the American tropics, spread in farms, grasslands, and open spaces curbing the growth of native herbaceous plants. Also known as a famine weed, *Parthenium* is an allelopathic plant, and its invasion results in reduced food grain yields. Its pollen grains have been found allergic to human beings.

Eichhornia, commonly called water hyacinth, is an aquatic plant, native to South America. It proliferates in freshwater ponds and rivers, dries up the lentic water systems, and clogs rivers. One of the fastest growing aquatic weeds, *Eichhornia* threatens the survival of many aquatic species.

Among the most cited examples of animal alien species is the Nile perch introduced in Lake Victoria of South Africa. Nile perch (*Lates niloticus*), a native to the Nile, Congo, Niger, Senegal, and many other river basins in the Afrotropical realm was introduced to Lake Victoria in the 1950s. This alien predatory fish threatened the entire freshwater lake ecosystem eliminating many species of endemic Cichlid fish. When this fish was eliminated, the Nile perch began preying upon small shrimp and minnows.

Hunting and Overexploitation

Hunting has been an old habit of human beings. Hunting is not just to fulfill essential needs or to take safety measures, but to satiate a will by posing as a triumphant. Killing animals by the "art" of hunting for taste, economic gains, pleasure, joy, recreation, and leisure management are the other objectives of human satiation. The two most striking examples of the consequences of hunting include the disappearance of the Cheetah (*Acinomyx jubatus*), India's fastest mammal, and the dodo (*Oidus impetus*), a unique bird of Mauritius. Over-exploitation of natural resources wantonly for meeting economic targets is throwing several species of plants and animals into terrestrial and aquatic ecosystems.

Poaching and Smuggling

The Trade of endangered wild animal products, such as tusks, horns, furs, etc. is banned. However, several animals in their wilderness are being killed for their products, and smuggling of the live specimen of endangered and threatened plant species, herbal products, etc. is going on unabatedly. Notorious poachers and smugglers manage illegal trade with the help of the mafia. These malpractices are posing a serious threat to many precious species of plants and animals.

Agriculture

Agriculture is the largest land-based human activity involving crop husbandry, forestry, horticulture, floriculture, animal husbandry, dairying, poultry, etc. that has devoured a good chunk of Earth's biodiversity. Agriculture is not just the art of making provision of foods, clothes, and most of the vital needs of humanity, but also an activity compromising with nature's biodiversity. The proportion of biodiversity used in managing agriculture is referred to as agrobiodiversity. Agrobiodiversity encompasses significant geographical space on the surface of the Earth but represents just the tip of the biodiversity.

Only a few plant species are grown for food production. Amongst them, only three plant species – wheat, rice, and corn – occupy the largest cultivated land and fulfill about three-fourths of human energy needs. We hardly encounter more than three dozen of plants of cereals, vegetables, fruits, beans, etc. providing us food. For fulfilling our clothing requirements, we depend on cotton plants almost exclusively. There are some other plant species too that are cultivated for human use, but the area occupied by these crops is very limited. Forestry involves only a limited number of trees and shrubs raised in monocultures. Animal husbandry involves only a few animal species: bovine, ovine, equine, canine, etc.

The worst fate agrobiodiversity faces is the depletion of genetic biodiversity. At some point in time, numerous varieties of food-providing plants used to be cultivated. In modern agriculture, the genetic base of food crops carries no value. The only criterion of cultivating a crop is productivity. In order of chasing the highest possible yields, only a few varieties of a crop are given room. These are the so-called high-yielding varieties (HYVs). In animal husbandry practices only a few breeds of cattle, buffaloes, goats, sheep, horses, camels, and poultry birds have been left on Earth. This extremely narrow genetic base has turned our agriculture into an extremely vulnerable sector. Narrow species and genetic bases invite more insect onslaughts and pest epidemics. This state of agrobiodiversity management imposes the use of deadly pesticides for plant protection and creates a trail of soil, water, air, and food pollution.

Environmental Pollution

Pollution has disruptive effects on all components of the environment. It plays a life-annihilating role in soil, water, and terrestrial ecosystems. Soil pollution due to frequent applications of pesticides in food crops and heavy metal pollutants in the air and consequently depositing on soil wipe out soil microflora and fauna. Pollutants in terrestrial ecosystems poison food chains and terminate sensitive species. Pollution of water bodies induces eutrophication and kills many aquatic species. The use of weedicides and herbicides kills many valuable plants – in the name of the weeds – not intended for cultivated lands. Oil spill changes the quality of sea habitats. Heavy metal pollution in water systems accumulates in the bodies of organisms and kills many sensitive organisms. Many predatory birds feeding on fish from polluted water bodies die due to biomagnification of the toxic pollutants.

Man-Wildlife Conflicts

Wild animals tend to migrate from one place to another. They are dynamic to the extent that they often break the boundaries of their habitats even if they face no dearth of food. However, in most cases, wild animals cross their boundaries when they face a shortage or lack of foods of their natural choice in their habitats. As a human intervention in wildlife habitats is increasingly intensified, a state of ecological disaster is ensuing to the extent that wild animals are forced to roam around human settlements and attack human beings and their livestock.

There are frequent cases of animal attacks on and killing of human beings and domestic animals in the settlements near protected areas. In response to their protection, local people also kill the animals. Thus, a man-wildlife conflict goes on unabatedly. Even the herbivorous animals that do not attack or harm people directly are killed because they destroy their crops.

When an animal gets injured by a man, it develops a tendency to retaliate. A weak and infirm animal also develops a liking to attack human beings. Females of animals attack human beings in a bid to protect their offspring. If a lion, tiger, leopard, or any other carnivore tastes human flesh, it becomes a man-eater. It is easy to kill a wild animal if it turns into a man-eater. But in the process of tracing a man-eater, many innocent carnivores become victims of the human fury of revenge.

There are also numerous cases of obstruction of wild animals' migratory routes by the construction of roads, tourist places, industrial areas, etc. As a result, animals tend to intrude on these settlements and often make humans easy prey for themselves. In this man-wildlife conflict, eventually, the wildlife suffers from heavy loss.

There is hardly a provision for fair compensation to farmers against the damage to their crops by wild animals. There are also no fool-proof ways of crop protection from wild animals. As a result, the local farmers resort to getting rid of the menace by killing wild animals.

Measures to Avert Man-Wildlife Conflicts

- If there is plentiful availability of foods, fodder, and water within protected areas and natural forest areas, chances are that the animals would not be compelled to cross the boundaries of their habitats.
- Farmers in the vicinity of protected areas may like to adopt alternative cropping patterns involving crops not palatable for wild herbivores. Thus, the animals would not be prompted to browse the crops they do not like.
- Migratory routes of the wild animals should not the disrupted or obstructed. Appropriate corridors should be provided to facilitate the seasonal migration of wild animals.
- Dense bio-fencing involving specific trees and bushes should be raised on the boundary of villages near national parks and wildlife sanctuaries. The same should also be done on the boundary of the protected areas.
- There should be the provision of adequate compensation to farmers whose crops are destroyed by herbivorous animals.

Extinction of Species

Extinction of the species is irreversible and, undoubtedly, the most serious issue about the biodiversity of the Earth. Once a species is gone forever from the face of the Earth, valuable information contained in its DNA is also gone. Once a species vanishes forever, we shall lose it at the cost of so many other species. Once a species goes extinct, chances of its further evolution are also lost forever. Once a species disappears forever, many other species dependent on it also come on the verge of extinction.

When is a species said to be extinct? When the last known member of the species anywhere in the world is dead. There are many members of many species which are no more alive in the wilderness but only in captivity, that is, under human-controlled conditions. Such species are said to be extinct in the wild. Under both these situations, the species would be regarded as globally extinct.

There is also a scenario representing a species but left in such a small measure that its impact on the community it exists within is almost negligible. Such a species is said to be ecologically extinct. For example, tigers in the wilderness are left in such a small number that their impact on the prey populations of herbivores is non-significant.

A climax species in an ecosystem plays a significant role in ecosystem functioning, such as biodiversity, soil and water conservation, microclimate development, ecosystem productivity, etc. Due to human intervention, however, the species is reduced in number to the extent that its phenomenal contribution to ecosystem functioning is practically ceased. Such a species is said to be functionally extinct and the situation is functional extinction.

Causes of Extinction

Species extinction is not a new phenomenon of life. The species have been done to extinction and new species have been evolving over the long evolutionary history of life on the living planet. Multicellular organisms evolved on Earth about 570 million years ago. According to some estimates, as many as 30 billion species have thrived on Earth during the entire history of biological evolution on the Living Planet. It is believed that 99.9% of all the species that evolved at some point in the long geological past have gone extinct. Extinction, in essence, is the ultimate fate of a species.

The natural extinction – that is, the extinction of a species in its millions of years-long journey of existence has to inevitably meet – has never been a matter of human worries. In recent decades, human activities resulting in environmental disruptions have led to accelerated extinction at an alarming rate. This is a worrisome state of human behavior with a variety of life that has been flowering on Earth. There are, on the whole, three processes by which the species are met to the fate of their extinction, viz. i) natural extinction, ii) mass extinction and iii) anthropogenic extinction.

Natural Extinction

All the individuals of all the species die as per the natural death program – the apoptosis – they are born with. Species, however, live for a pretty long period of natural evolution, but not forever. An individual species has also to die in response to some laws of life not known to date. The death of a species with all its genotypes and individual members from everywhere on Earth in a natural way is what is called natural extinction. It is generally thought that some species disappear and the others 'more adapted to changed environmental conditions' replace them. Singh (2019), in his theory of *Fertilizing the Universe*, ascribes this phenomenon to the role light-converted-to-biochemical energy plays in living energy systems. The natural extinction that has occurred in the geological past is also referred to as background extinction.

Mass Extinction

At various points of the geological history over various geological times extinction of a large number of species due to natural global catastrophes is supposed to have taken place. Mass extinction means the rate of species disappearance is significantly higher than that of background extinction. The mass extinction has taken place over millions of years. According to some estimates, more than 65% of all species met the fate of their extinction in global catastrophic events over a few millions of years, a period much smaller compared to the history of the Earth spanning over some 4.6 billion years.

Scientists believe that five mass extinction events have occurred on Earth over the past 500 million years leading to dramatic alterations in the characters of ecological communities. A mass extinction happens to be a big jolt to the Earth's natural biodiversity. It is believed that it took some 50 to 100 million years for the global biodiversity to rejuvenate and establish following an event of mass extinction.

The most severe mass extinction resulting in the disappearance of as many as 95% of marine species is supposed to have occurred about 225 million years ago. The most talked about and the last mass extinction that wiped out the giant dinosaurs ruling over the earth for as many as 140 million years took place some 65 million years ago.

The global catastrophic events that led to mass extinction include both environmental and biological factors. These are global cooling, a decrease in sea level, predation, and competition. Various epochs of global cooling brought the earth's temperature down to a level where most of the species could not maintain their physiology and vanished. A fall in sea levels could have been attributable to global cooling. Most of the evaporated water froze as ice and snow especially on the poles and in the mountains leading to a significant fall in the sea levels. Unusual rates of predation and competition among species might have resulted due to ecological changes induced by adverse thermodynamic conditions of the biosphere. One of the recently emerged theories of mass extinction, especially of the dinosaurs, suggests that a giant comet or an asteroid hit the earth leading to wiping out masses of species directly and due to huge clouds of dust preventing sun rays from reaching the earth for a long time.

Anthropogenic Extinction

Wiping out living species due to human actions or human-made circumstances is referred to as anthropogenic extinction. This is the worst kind of extinction at the hands of a single species – the human species – disrupting ecological systems and ecological processes, over-exploiting natural resources, and annihilating living species that are not of direct consumptive value and/ or are thought to be undesirable or harmful. It is thanks to human activities that the present rates of extinction are 1000 to 10,000 times higher than the background extinction.

According to IUCN, about 4500 species of mammals, birds, and amphibians are deemed at risk of extinction. These include 1200 mammalian species, 1300 avian species, and 2000 amphibian species. Of all the described species, the proportion of the species deemed to go extinct is 26% mammalian, 13% avian, and 41% amphibian species. It is also estimated that with the current rate of species loss, we may witness the loss of nearly half of all living species by 2100.

Susceptibility to Extinction

All living species are not equally vulnerable to the processes leading to extinction. Some specific characteristics make some species relatively more susceptible. These are:

1. **Feeding at a high trophic level:** Top carnivores are left with a minimum proportion of energy trapped by producers in an ecosystem to sustain with. Herbivores, closest to producers, on the other hand, have plentiful

energy. In other words, the further from the photosynthesis a species is, the more susceptible to extinction it is. The Bengal tiger and the Bald eagle are the main examples facing a high degree of susceptibility to extinction because of top carnivory.

2. **Small population size and low fecundity:** These characteristics hold a species in the state of continuous pressure of competition from the species with larger population sizes and high fecundity. Blue whales and Giant Panda are peculiar examples facing this kind of vulnerability.

3. **The localized and narrow range of distribution:** The smaller the geographical area and narrower the range of distribution the greater the chances of a species to stay with vulnerability to extinction. The island species and the woodland caribou are examples.

4. **Fixed migratory routes:** Species following fixed migratory routes, such as the Blue whale and the Whooping crane, are likely to be at a greater loss as they may encounter several obstructions imposed by adverse environmental conditions and increased chances of becoming victims of predatory species.

5. **Large body size:** Species like Blue whales, elephants, Bengal tigers, lion are more likely to face problems of ecological space and meeting food requirements proportionate to their body requirements.

IUCN Red List

The IUCN Red List or Red Data List of threatened species is the world's most comprehensive inventory of biological species. Founded by International Union for Conservation of Nature (IUCN) in 1964, the Red Data List comprises the taxa faced with the risk of extinction which is relevant to all species and all regions of the world. The main objectives of the IUCN Red List include:

i) Identification and documentation of endangered species

ii) Provision of a global index of the decline of biodiversity

iii) Development of awareness about the importance of threatened biodiversity

iv) Defining conservation priorities at the local level and guiding conservation action

There are nine Red List categories of the species defined by IUCN (2019) in accordance with the degree to which they face threat to extinction. Each taxon of the biosphere, except microorganisms, can be classified into these categories listed and defined in Table 5.1.

Table 5.1: The IUCN Red List Categories

Red List Category	A taxon is placed in the category when:
Extinct (EX)	...there is no reasonable doubt that the last individual has died.
Extinct in the Wild (EW)	...when it is known to survive only in cultivation, captivity, or as a naturalized population well outside the past range; exhaustive surveys in known and/ or expected habitats fail to record the presence of an individual.
Critically Endangered (CR)	...it is facing an extremely high risk of extinction in the wild in the immediate future.
Endangered (EN)	...it is nor Critically Endangered but is facing a very high risk of extinction in the wild in the near future.
Vulnerable (VU)	...it is not Critically Endangered or Endangered, but is facing a high risk of extinction in the wild in the medium-term future
Near Threatened (NT)	...it is evaluated against the criteria but does not qualify for CR, EN, or VU now, but is close to qualifying for or is likely to qualify for a threatened category in near future
Least Concern (LC)	...it has been evaluated against the criteria but is not found qualifying for CR, EN, VU, or NT (indicative of widespread and abundant taxa)
Data Deficient (DD)	...there is inadequate information to make a direct or indirect, assessment of its risk of extinction.
Not Evaluated (NE)	...it has not yet been assessed.

Source: IUCN (2019)

Individual countries and/ or organizations also prepare the Regional Red Lists that are helpful in assessing the risk of extinction to species within different political management systems.

Summary

A large number of genetic resources, species, and ecosystems are under severe threat. The major factors becoming threats to biodiversity are anthropogenic. The primary threat to biodiversity lies in habitat destruction. Almost all the habitats have been modified to suit the requirement of the human species. What used to be habitats of a variety of wild animals are now human settlements – villages, towns, cities, commercial and industrial establishments, playgrounds, tourist spots, honeymoon huts, and the likes. Habitat fragmentation, i.e. dividing a large intact habitat into two or multiple fragments, poses threat to biodiversity in many ways, viz. i) by limiting dispersal and colonization potential of species; ii) by reducing the foraging ability of the wild animals; iii) by disturbing ecological niches and/ or separating species from their niches. The exotic species exert pressure on a new community in several ways and may lead to serious ecological and environmental consequences. The ecological effects of exotic species on island ecosystems – that are home to most of the world's rare and threatened species – are especially grave. Several animals in their wilderness are being hunted and killed for their products, and smuggling of the live specimen of endangered and threatened plant species, herbal products, etc. is going on unabatedly. Agriculture is the largest land-based human activity that has devoured a good chunk of Earth's biodiversity. Pollution has to disrupt effects on all components of the environment.

As a human intervention in wildlife habitats is increasingly intensified, a state of ecological disaster is ensuing to the extent that wild animals are forced to roam around human settlements and attack human beings and their livestock. In response to their protection, local people also kill the animals. Thus, a man-wildlife conflict goes on unabatedly. There are also numerous cases of obstruction of wild animals' migratory routes by the construction of roads, tourist places, industrial areas, etc. In this man-wildlife conflict, eventually, the wildlife suffers from heavy loss.

Extinction of the species is irreversible and, undoubtedly, the most serious issue about the biodiversity of the Earth. Once a species goes extinct, chances of its further evolution are also lost forever. Once a species disappears, many other species dependent on it also come on the verge of extinction. When the last known member of the species anywhere in the world is dead, it is said to be extinct. There is also a scenario representing a species but left in such a small measure that its impact on the community it exists within is almost negligible. Such a species is said to be ecologically extinct. Due to human intervention, a species is reduced in number to the extent that its phenomenal contribution to ecosystem functioning is practically ceased. Such a species is said to be functionally extinct and the situation is functional extinction.

Species extinction is not a new phenomenon of life. It is believed that 99.9% of all the species that evolved at some point in the long geological past have gone extinct. The natural extinction – that is, the extinction of a species in its millions of years-long journey of existence has to inevitably meet – has never been a matter of human worries. In recent decades, human activities resulting in environmental disruptions have led to accelerated extinction at an alarming rate. There are three processes by which the species are met with the fate of their extinction: i) natural extinction, ii) mass extinction, and iii) anthropogenic extinction. Mass extinction means the rate of species' disappearance is significantly higher than that of the background extinction. Five mass extinction events have occurred on Earth over the past 500 million years leading to dramatic alterations in the characters of biotic communities. The Sixth Mass Extinction is said to be going on. Wiping out living species due to human actions or human-made circumstances is referred to as anthropogenic extinction. This is the worst kind of extinction disrupting ecological systems and ecological processes. It is thanks to human activities that the present rates of extinction are 1000 to 10,000 times higher than the background extinction.

According to IUCN, about 4500 species of mammals, birds, and amphibians are deemed at risk of extinction. It is also estimated that with the current rate of species loss, we may witness nearly half of all the living species lost by 2100. Some specific characteristics make some species relatively more susceptible. These are: feeding at a high trophic level, small population size and low fecundity, localized and narrow range of distribution, fixed migratory routes, and large body size.

The IUCN Red List or Red Data List of threatened species is the world's most comprehensive inventory of biological species. The main objectives of the IUCN Red List are: identification and documentation of endangered species, provision of a global index of the decline of biodiversity, development of awareness about the importance of threatened biodiversity, defining conservation priorities at the local level, and guiding conservation action. There are eight Red List categories of the species defined by IUCN following the degree to which they face the threat of extinction. These are extinct, extinct in wild, critically endangered, endangered, vulnerable, lower risk, data deficient, and not evaluated.

Exercises

A. Multiple Choice Questions

1. How many life forms that ever existed on Earth during the history of evolution are considered to have gone extinct?

   a. 1% b. 50%

   c. 90% d. 99.9%

2. Which of the following is the right statement?

   a. A species at a certain point of time in its evolutionary history is destined to vanish.

   b. There is hardly any threat to biodiversity due to natural factors except those inherent in the designs of natural evolution.

c. Agriculture has also been responsible for biodiversity erosion.
d. All of the above statements are right.

3. The K-T Extinction Event is about the extinction of the race of
   a. Dinosaurs
   b. Tasmanian Tigers
   c. Saber Tooth Tigers
   d. Woolly Mammoths
4. Which of the following is the case of the most recent extinction?
   a. Dinosaurs
   b. Archaeopteryx
   c. Woolly Mammoths
   d. Dodo
5. "Dead as __________." This case of extinction is often referred to this way. What is this?
   a. Dead as Dinosaurs
   b. Dead as Donkey
   c. Dead as Archaeopteryx
   d. Dead as Dodo
6. The Indian Bustard is counted as an endangered species due to
   a. deforestation in the Thar Desert
   b. Expansion of the Thar Desert
   c. Hunting
   d. Climate change
7. The communities most vulnerable to exotic species are
   a. Oceanic islands
   b. Tropical rainforests
   c. Temperate forests
   d. Mangroves
8. What of the following is the cause of biodiversity loss?
   a. Habitat fragmentation
   b. Introduction of exotic species
   c. Overexploitation of forests
   d. All of the above
9. The example of the Nile perch (*Lates niloticus*) pertains to
   a. An exotic plant introduced in Nile River
   b. A native fish of the river Nile introduced in Lake Victoria
   c. An exotic fish introduced in the river Nile
   d. An exotic weed introduced in the Lake Superior

10. Which of the following agricultural practices contributes to biodiversity loss?

    a. Applications of weedicides and herbicides

    b. Raising monocultures

    c. Sowing high-yielding varieties

    d. All of the above

11. An extinct species is

    a. The last known member of the species anywhere in the world which is dead

    b. No more alive in the wilderness but only in the captivity

    c. Kept under human-controlled conditions

    d. The one that can vanish at any time

12. The natural extinction that has occurred in the geological past is also referred to as

    a. Extinction in the wild

    b. Threat of extinction

    c. Background extinction

    d. Mass extinction

13. What is true about anthropogenic extinction?

    a. It is the worst kind of extinction occurring at the hands of the human species.

    b. The rates of extinction are 1000 to 10,000 times higher than the background extinction.

    c. It is accelerated due to accelerated ecological disruptions.

    d. It is on account of the facts mentioned at a, b, and c above

14. Susceptibility to extinction is on account of

    a. Feeding at a high trophic level

    b. Small population size and low fecundity

    c. The localized and narrow range of distribution

    d. All of the above

15. Which of the following species is likely to face a higher probability of extinction?

    a. The oak, a producer
    b. The zebra, a herbivore
    c. The wolf, a first order carnivore
    d. The lion, a top carnivore

16. Species like Blue whale, elephant, Bengal tiger, and lion are more likely to face higher chances of extinction because

    a. They all are herbivores
    b. They all are carnivores
    c. Of their larger size
    d. They are more susceptible to diseases

17. What is not one of the main objectives of the IUCN Red List?

    a. Identification and documentation of endangered species
    b. Development of awareness about the importance of threatened biodiversity
    c. *Ex-situ* conservation of the threatened species
    d. Provision of a global index of the decline of biodiversity

18. How many Red List categories of the species defined by IUCN as per the degree to which they face extinction threats are there?

    a. Eight
    b. Ten
    c. Sixteen
    d. Twenty

19. The first and the last category of the IUCN Red List, respectively, are

    a. Extinct, Not Evaluated
    b. Not Evaluated, Extinct
    c. Endangered, Vulnerable
    d. Lower Risk, Data Deficient

20. A species is facing an extremely high risk of extinction in the wild in the immediate future. In the IUCN Red List, it will be put in the category of

    a. Extinct in the Wild
    b. Critically Endangered
    c. Endangered
    d. Vulnerable

Answers: 1. **d**, 2. **d**, 3. **a**, 4. **d**, 5. **d**, 6. **c**, 7. **a**, 8. **d**, 9. **b**, 10. **d**, 11. **a**, 12. **c**, 13. **d**, 14. **d**, 15. **d**, 16. **c**, 17. **c**, 18. **a**, 19. **a**, 20. **b**

B. Short-Answer Questions

1. List various threats to biodiversity.
2. What do you mean by habitat fragmentation?
3. Name some exotic animal species that have destroyed endemic fauna.
4. Give some examples of habitat destruction due to introduced plant species.
5. How does hunting destroy a habitat?
6. Why are the top carnivores regarded as keystone species?
7. How does environmental pollution affect biodiversity?
8. When is a species said to be extinct?
9. Write the objectives of the IUCN Red Data List.
10. When is a species said to be Extinct in the Wild?

C. Long-Answer Questions

1. What is meant by the extinction of species? What are the various causes of species extinction?
2. What could be the consequences of species extinction?
3. Give an account of the main causes leading to the destruction of natural habitats.
4. Describe with examples the ecological effects of exotic species.
5. Write short notes on the following:
   (a) Man-Animal Conflicts
   (b) Susceptibility to Extinction
   (c) Red Data List

References

IUCN. 2019. Guidelines for Using the IUCN Red List Categories and Criteria: Version 14. Gland, Switzerland and Cambridge, UK: IUCN. 113 pp.

Singh, Vir. 2019. Fertilizing the Universe: A New Chapter of Unfolding Evolution. London: Cambridge Scholars Publishing. 285pp.

WWF. 2020. Living Planet Report 2020 – Bending the Curve of Biodiversity Loss. Almond, R. E. A., Groten, M. and Petersen, T. (eds). Gland, Switzerland: WWF.

Web Resources

https://blog.nationalgeographic.org/2019/09/23/global-biodiversity-is-in-crisis-but-there-is-hope-for-recovery/
https://earthjustice.org/features/biodiversity-crisis
https://pioneeredge.in/is-death-of-the-earth-due/
https://www.bbc.com/news/science-environment-54357899
https://www.huffpost.com/entry/un-environment-programme-_n_684562
https://www.wired.com/story/2021-biodiversity-crisis/

6

Biodiversity Conservation

Most of the ecosystems prevailing on Earth have ceased to be virgin. They are now modified following anthropogenic designs. The Blossoming of a community in its climax stage is a rare sight. As human intervention with nature has gone on increasing, all types of biodiversity – genetic, species and community, and alpha, beta, and gamma – have been put under stress. The Earth has already lost much of its diversity due to anthropogenic factors and now it seems to be under severe stress due to climate change, which is also largely due to human-induced environmental disruptions.

Continuous loss of biodiversity at an alarming rate is posing a serious threat to life on earth. With this dismal state of the environment, we cannot usher in a sustainable future. We do not have the right to snatch the future from our younger generations. The Earth belongs to our children more than it does to us. The Earth belongs to future generations more than it does to present generations. Conservation of biodiversity is a matter of ethics. Ethics has the human face of biodiversity. Economic attributes of biodiversity have importance, but ethics of life should not be ignored.

How would the nature of the Earth look like without biodiversity? It won't look beautiful. Beauty is an expression of a life-supporting and life-enhancing environment. Beauty emanates from the richness of life, not from lifelessness. Embodying richness of life, biodiversity, thus, spells out its aesthetic values. The future generations inherit the right to live amidst a beautiful environment of blossoming beauty. Conservation of biodiversity – thus – is a must for the sake of ethical and aesthetic values.

Biodiversity has not been understood in its entirety. Nor does it seem to be possible. Many of the species have disappeared as they were supposed to be of any economic value. Several undesirable plants referred to as "weeds" vanished due to human activities. But how can one know these plants are not valueless? These may be absorbing some optional values which we could not capture to date, and those optional values might prove vital for us in the future. So, species must be saved and conserved for preserving the optional values that may open new windows of our knowledge about the plants.

No species on Earth is useless. Every species came into being due to natural evolution. How can then a species be useless or valueless? Every taxon in existence is laden with valuable existence value. Existence value is a vital attribute of natural evolution. Biodiversity must be conserved for recognizing and upholding the sacred existence value of all life forms, and all species and for appreciating natural evolution and marveling at its living attributes adorning the planet with a diversity of life. The beauty of the planet lies in its biodiversity. In other words, biodiversity is the natural foundation of the beauty our Earth embodies.

The bloom of biodiversity on Earth contributes to maintaining biogeochemical cycles, water cycles, weather cycles, seasons, and climates, and strikes an ecological balance vital for the biosphere to effectively maintain its life-giving, life-supporting, and life-enhancing functions (Singh 2019).

We are bound with moral obligations to protect, regenerate and increase natural resources with the richness of biodiversity. The conservation decisions we take today would have a phenomenal impact on the biodiversity scenario of the future.

Strategies for Biodiversity Conservation

Biodiversity conservation means adopting and implementing sound management practices so that all living species and their genotypes are held in natural and/ or artificial systems for their sustainable use by the present and future generations. The first and foremost measure of biodiversity conservation is the prevention of further deterioration, degradation, depletion, and degeneration of natural habitats of the species. There are two major challenges in implementing conservation strategies: limited space and human pressure on natural resources. Two basic strategies of biodiversity conservation applied in the world are: in-situ conservation and ex-situ conservation.

In-Situ Conservation Strategies

Conservation of the biological diversity in its natural habitats is what the in-situ conservation strategy refers to. All the species prospering on Earth have evolved in the natural environments which were appropriate for their evolution and where they could adapt along with other associated species in tune with the operating environmental factors conducive to their survival, functions, and reproductive activities. Giving due protection to the entire ecosystem where communities of diverse species prosper is what the in-situ conservation approach calls for. For the operationalization of this strategy, a group of specific ecosystems is protected and thus a network of protected areas is earmarked for biodiversity conservation.

Protected Areas

Protected areas include earmarked and defined geographical areas specially dedicated to the protection and maintenance of biodiversity in natural environs or wilderness. The range of conservation extends from the protection of the populations of specific species to the preservation of the entire ecosystem. For the protection of the species within the wilderness, the protection of the wilderness itself is necessary. The associated natural and cultural resource conservation processes are also taken care of in the protected areas. The protected areas include national parks and wildlife sanctuaries and are protected and managed by employing legal measures and/ or by other effective means. Larger areas are preserved with the objective that biodiversity continuously evolves and flourishes along with the ecosystem it inhabits. The larger wild animals, like elephants, need larger reserves for the provision of adequate ecological space for their frequent movement. The larger the protected area the less vulnerable to disturbance it is.

A protected area is aimed at giving due protection aimed at conserving nature's biodiversity as well as geo-morphological, ecological, and biological features, ecological processes, and various social, cultural, ethical, and aesthetic values. It constitutes an internationally recognized system for categorizing a management approach adopted in individual areas set aside for the purpose. IUCN has developed strategies for developing protected areas for biodiversity conservation at local, national, sub-national, and international bases. These protected areas are supported and governed by local and national governments and even by private organizations and individuals. Nomenclature, governance system, management approach, etc. relating to the protected areas generally vary from place to place and from country to country.

The management approaches might be different, but the management objectives need to be well-defined and fulfilled. To categorize the protected areas following their underlying management objectives, the IUCN has developed a system that has been recognized by prominent international organizations, including UNO, and by several national governments. Even the Convention on Biological Diversity (CBD) ratified at the Earth Summit 1992, in its Goal 43, Paragraph 4.3.7 of the Programme of Work on Protected Areas, calls for using the IUCN system of protected area management categories. The IUCN management categories have become a worldwide standard for the planning, establishment, and management of protected areas. These categories and their salient features are presented in Table 6.1.

Table 6.1: IUCN Categories of Protected Area Management and their brief description

IUCN Category	Description
Ia – Strict Nature Reserve	Strictly protected areas set aside to protect biodiversity, including, as far as possible, the geological and geomorphological features; human interference is strictly controlled
Ib – Wilderness Area	Large unmodified or slightly modified areas retaining their natural features; no permanent or significant human habitation
II – National Park	Large natural or near natural areas set aside to protect large-scale ecological processes, species and ecosystem characteristics; provision of educational, scientific, spiritual and recreation opportunities
III – National Monument or Feature	Areas set aside for protecting a specific natural monument, for example, a geological feature like a grove, a landform, sea mount etc.
IV – Habitat/ Species Management Area	Emphasis on the protection of particular species or habitats
V – Protected Landscape/ Seascape	Protection of the areas of distinctive characteristics (e.g., significant ecological, biological, cultural and scenic value) created thanks to people's interactions or specific natural factors, as well as integrity of such interaction
VI – Protected Area with Sustainable Use of Natural Resources	Large area, mostly maintaining natural conditions, with a proportion under sustainable natural resource management, set aside for protection and biodiversity conservation

Source: Compiled from https://www.biodiversitya-z.org/content/iucn-protected-area-management- categories.pdf

There are as many as 202,467 protected areas throughout the world as of October 2017 according to the World Database on Protected Areas (WDPA). The world's protected areas cover about 200,000,000 km^2 area or 14.7% of the world's land, Antarctica excluded. The Convention on Biodiversity under the Aichi Biodiversity Targets has set 17% area to be covered for biodiversity conservation in 2020. Thus, there is still a gap of 2.3% to achieving the target of protected areas. Apart from the land area, some 10% of the territorial waters are also protected.

Not all countries of the world have an equal proportion of land under protected areas. Protected areas in Afghanistan, Libya, Turkey, Lesotho, Mauritania, Syria, Yemen, Somalia, Barbados, and Bosnia and Herzegovina are almost negligible, ranging from only 0.10% in Afghanistan to 1.40% in Bosnia and Herzegovina, as per the scenario of 2018. On the other hand, some countries, notably New Caledonia, Venezuela, Slovenia, Bhutan, Brunei, Seychelles,

Hong Kong SAR, China, Greenland, Luxemburg, and Congo have a very large proportion of their lands under protected areas, ranging from 54.40% in New Caledonia to 40.74% in Congo, as per the scenario of 2018. Ten countries with minimum and 10 with a maximum percentage of the protected areas are shown in Fig. 6.1 and Fig. 6.2, respectively.

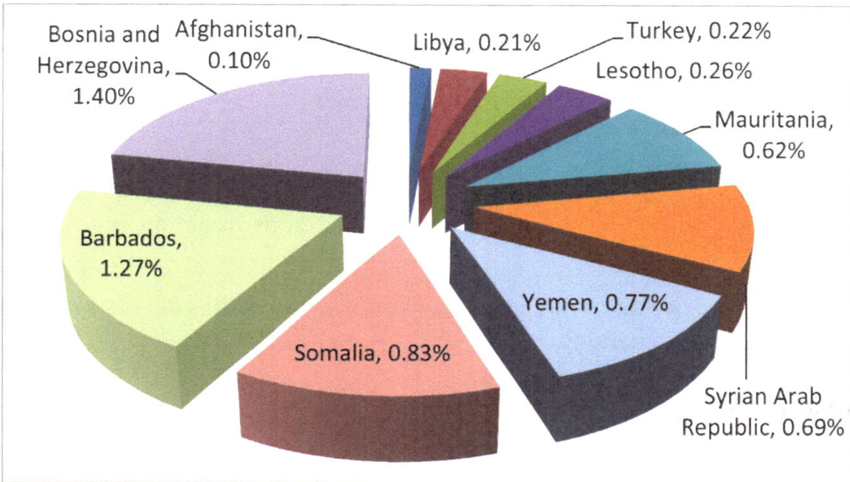

Fig. 6.1: Ten countries with minimum (almost negligible) coverage of terrestrial protected areas (Source: Based on the information in World Database on Protected Areas (WDPA)

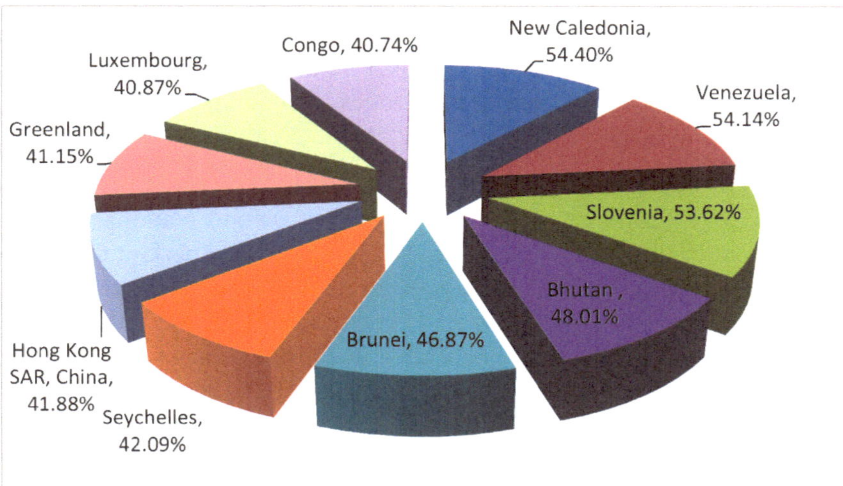

Fig. 6.2: Ten countries with maximum percentage of the terrestrial protected areas
Source: Based on the information in World Database on Protected Areas (WDPA)

A protected area under the in-situ conservation strategy is ecologically more complex, healthier, more vibrant, more resilient, more productive and more sustainable than an anthropogenic one managed under ex-situ conservation strategy. These protected ecosystems play pivotal role in building up their own microclimate and phenomenally contribute to maintain macroclimate. Some other vital attributes of the protected areas are as follows:

i) Conservation and maintenance of native populations of the species and sub-species;

ii) Maintenance of the communities' size;

iii) Conservation of genetic diversity of all the species;

iv) Prevention of introduction of alien species, thus avoiding invasion of non-native species;

v) Making possibility for species/ habitats to shift in response to environmental changes.

National Parks

A national park is meant for the welfare of wildlife. Human activities like agriculture, forestry, grazing or land use change are not allowed. A national park is owned by a sovereign state and can include a reserve of natural, semi-natural or developed area. People also cannot enjoy private ownership rights. Hunting, exploitation of resources and habitat manipulation are strictly prohibited in a national park. In 1969, IUCN did not fix area size to a national park and, apart from other conditions, only referred to it as a 'relatively large area'. In 1971, a minimum size of 1,000 hectares 'within zones in which protection of nature takes precedence' was added to its defining characteristics.

There are 6,555 national parks throughout the world that meet the criterion defined by IUCN. Established in 1872, the Yellowstone National Park of USA has the credit of being the first and the oldest national park in the world. However, when it was established, it was not designated as a National Park. The credit of the oldest legally protected area should go to the Tobago Main Ridge Forest Reserve which was established in 1776, that is nearly 100 years before the Yellowstone National Park. The other much older protected area established in 1778 was Bogd Khan Uul Mountain. Some interesting features relating to the world's national parks are presented in Table 6.2. India in its territory embraces some of the world's robust national parks located in diverse geographical areas ranging from the desert to the Himalayan mountains. The ten such parks with the largest area are listed in Table 6.3.

Table 6.2: Interesting features of the world's national parks

The oldest protected area	The Tobago Main Ridge Forest Reserve established in 1776
The continent with most of the National Parks	Asia
The country with most of the National Parks	Australia, 285 National parks
Ten oldest National Parks	Bogd Khan Uul National Park, Mongolia (1778) Yellowstone National Park, USA (1872) Royal National Park, Australia (1879) Banff National Park, Canada (1885) Glacier National Park, Canada (1886) Yoho National Park, Canada (1886) Tongariro National Park, New Zealand (1887) Yosemite National Park, USA (1890) Kings Canyon National Park, USA (1890) Sequoia National Park, USA (1890)
The largest National Park	Northeast Greenland National Park in Denmark, size 972,000 km^2 (more than 24 million acres); Wrangell – St. Elias in Alaska, size 8 million acres (32,375 km^2)
The smallest National Park	The Moyenne Island National Park spread over just 22 acres in Seychelles, designated National Park by Government of Seychelles in 2008; Gateway Arch National Park in Missouri spread over 192.83 acres (0.7804 km^2)
The biggest National Park of India	Hemis National Park spanning over 4,400 km^2
The oldest National Park in India	Jim Corbett National Park (1936)

Table 6.3: The ten largest national parks in India

Rank	Name	State	Area (in km^2)
1.	Hemis National Park	Jammu and Kashmir	4,400.0
2.	Desert National Park	Rajasthan	3,162.0
3.	Gangotri National Park	Uttarakhand	2,390.0
4.	Namdapha National Park	Arunachal Pradesh	1,985.2
5.	Khangchendzonga National Park	Sikkim	1,784.0
6.	Guru Ghasidas (Sanjay) National Park	Chhattisgarh	1,440.7
7.	Gir Forest National Park	Gujarat	1,412.0
8.	Sundarbans National Park	West Bengal	1,330.1
9.	Jim Corbett National Park	Uttarakhand	1,318.5
10.	Indravati National Park	Chhattisgarh	1,258.4

Wildlife Sanctuaries

A sanctuary is meant to be an area legally protected or reserved only for the conservation of wild animals. Habitat of wild animals and their surroundings are given due protection for their safety, comfortable living, free movement and breeding. Poaching, capturing and killing of animals in a sanctuary are strictly prohibited. There are certain human activities such as harvesting of minor forest products and timber may be provided and private ownership rights may be given provided the wildlife is not affected adversely.

Some of the world's biggest wildlife sanctuaries giving protection to wild animals in terrestrial and marine habitats are presented in Table 6.4. Natural Park of the Coral Sea encompassing an area of 129,49,94 km^2 and Pacific Remote Islands Marine National Monument with an area of 1,269,094 km^2 are among the world's largest animal wildlife sanctuaries.

In India the wildlife sanctuaries are classified as IUCN Category IV protected areas. There are as many 543 wildlife sanctuaries in India covering an area of 118,918 km^2 (Table 6.5, Fig. 6.3). Among these sanctuaries, 50 are governed by the Project Tiger aimed at conserving the Bengal tiger.

Table 6.4: Some of the world's largest wildlife sanctuaries

Wildlife sanctuary	Geographic location	Special features
Natural Park of the Coral Sea Area: 129,49,94 km^2 Established in: 2014	About 1200 km east of Australia encompassing New Caledonia; the Lagoons of New Caledonia, and the New Caledonia Barrier Reef – the world's second largest double barrier reef	Protects native wildlife, e.g. 25 species of marine mammals including manatee-like dugongs, 48 shark species, five sea turtle species and 19 species of nesting birds; one of the world's seven natural wonders – the Great Barrier Reef – is also given adequate protection
Kavango-Zambezi Transfrontier Conservation Area Area: 517,998 km^2 Established in: 2011	Around the convergence of the borders of Angola, Zambia, Namibia, Botswana, and Zimbabwe; covers Victoria Falls and the Okavango Delta; includes 35 national parks, forest reserves, games reserves, wildlife management areas	Protects Africa's largest contiguous population of elephants, cheetah, hippopotamus, African wild dogs, more than 600 bird species
Northeast Greenland National Park Area: 971,246 km^2 Established in: 1974	Extending across half of the Greenland; an international biosphere reserve as many as 77 times larger than the Yellowstone National Park	An international biosphere reserve; largely remote and inaccessible; provides protection to vast Arctic wilderness: e.g., polar bears, musk-oxen, arctic foxes and hares, giant walruses, four seal species etc.

Pacific Remote Islands Marine National Monument Area: 1,269,094 km^2 Established in: 2009, Expanded in: 2014	In the west of the Hawaiian Islands, managed by US Fish and Wildlife Service	Comprises islands, reefs and atolls; protection provided to more than 20 marine mammalians (the whales and the dolphins), invertebrate mollusks, crustaceans, sea turtles, world's largest arthropod (the coconut crab), about 200 coral species, sea birds etc.
Papahānaumokuākea Marine National Monument Area: 362,600 km^2 Established in: 2006 Expanded in: 2016	Encompasses the Hawaiian Islands National Wildlife Refuge and Midway Atoll National Wildlife Refuge; comprises 10 10 islands and atolls of the Northwestern Hawaiian Islands; larger than all the national parks in America and also than that of whole Germany; managed by National Oceanic and Atmospheric Administration (NOAA) and Interior's Fish and Wildlife Service (FWS)	Largest marine protected area in the world, provides refuge to about 7000 animal species, including the green sea turtle, an endangered species and the Hawaiian monk seal

Fig. 6.3: Wildlife sanctuaries in India (https://en.wikipedia.org/wiki/List_of_wildlife_sanctuaries_of_India)

Table 6.5: Some important wildlife sanctuaries in India

Name of location	Area in km^2	Important animals
Karakoram (Nubra Shyok) Wildlife Sanctuary, Ladakh	5000	Tibetan antelope, wild yak, leopards, Himalayan mouse, lynx, tufted duck, booted eagle, golden eagle, grey plover
Wild Ass Wildlife Sanctuary, Gujarat	4953.71	Ass, nilgai bustard, dalmatian pelican, hawks, harriers, falcons, Indian gazelle, desert fox (Indian and white-footed), jackals, African lynx
Changthang Wildlife Sanctuary, Ladakh	4000	Wild yak, Tibetan wolf, brown bear, bharal, mormot
Nagarjuna Sagar-Srisailam Wildlife Sanctuary, Telangana	3568	Tiger, panther, black bucks, mouse deer, sambhar, spotted deer, jackals, wild boar, wild bear, sloth bear, leopard, Indian giant squirrels, crocodiles, python, cobra
Hastinapur Wildlife Sanctuary, Uttar Pradesh	2073	Ganga River dolphin, swamp deer, smooth-coated otter, short-toed snake eagle, Egyptian vulture, white-eyed buzzard, Indian grey hornbill
Gir Wildlife Sanctuary, Gujarat	1153.42	Asiatic lion, Indian leopard, Striped hyena, four-horned antelope, wild boar. Mugger crocodile, Indian cobra, tortoise, monitor lizard
Chilka Lake Bird Sanctuary, Odisha	990	Water fowl, duck, crane, sand piper, golden plover, flamingoes
Kedarnath Wildlife Sanctuary, Uttarakhand	975.2	Himalayan black bear, Indian jackal, red fox, Himalayan monal, snow partridge, grey-cheeked warbler, Himalayan pit viper
Annamalai Sanctuary, Coimbatore, Tamil Nadu	958	Tiger, panther, elephant, spotted deer, sloth bear, sambhar, barking deer, wild dog
Manas Wildlife Sanctuary, Assam	950	Tiger, panther, wild boar, rhino, wild buffalo, golden languor, swamp deer, wild dog
Bhimbandh Wildlife Sanctuary, Bihar	681.99	Tiger, leopard, wild bear, nilgai, cheetal, barking deer, langoor, Indian hare, pea fowl, van murgi
National Chambal Wildlife Sanctuary, Uttar Pradesh	635	Ganga River dolphin, gharial, striped hyena, Indian wolf, red-crowned roof turtle, mugger crocodile, smooth-coated otter
Achanakmar Wildlife Sanctuary, Chhattisgarh	551.55	Indian leopard, sloth bear, wild boar, blackbuck, sambhar, nilgai, chital, striped hyena, Indian jackal, four-horned antelope
Balaram Ambaji Wildlife Sanctuary, Gujarat	542.08	Indian leopard, sloth bear, nilgai, striped hyena, Indian fox, Indian civet, Indian porcupine, Indian pangolin
Abohar Wildlife Sanctuary, Punjab	186.5	Black buck, wild boar, blue bull, Porcupine, black duck
Interview Island Wildlife Sanctuary, Andaman and Nicobar	133.87	Feral elephants, spotted dear, wild pig, three-stripped palm squirrel
Keoladeo Ghana Bird Sanctuary, Rajasthan	28.73	Siberian crane, storks, egrets, herons, spoon bill, spotted deer, black buck, wild boar, blue bull

Biosphere Reserves

The concept of a biosphere reserve was evolved in 1975 by the Man and Biosphere (MAB) program of UNESCO. This concept led to the creation of a special category of protected areas – land or coastal environments – with people as an integral component of the system. The following conservation, utilization, and socio-economic philosophy emerge out of the concept of biosphere reserves:

- In-situ conservation of all forms of life along with its support system
- Reconciling biodiversity conservation with its sustainable use
- Learning areas for sustainable development under diverse ecological, socio-cultural, and economic contexts
- Sites for testing interdisciplinary approaches to natural resource management
- Ecosystem management provides an opportunity for local people's participation in resource management
- Benefit sharing with local populations
- Ecological spaces for resolving conflicts between the public system and local people relating to biodiversity management (conservation and utilization)
- Provision of local solutions to global problems

At present, there are 701 biosphere reserves in the world within the boundaries of 124 countries. These include 21 transboundary sites belonging to the World Network of Biosphere Reserves, an organization promoting international cooperation by employing exchanging experience and technical know-how, capacity building, and promotion of best practices for the management of biosphere reserves.

The planning and management of biosphere reserves involve local communities as well as other interested stakeholders. Their role leads to three main functions as shown in Fig. 6.4. The natural resource management focusing on biodiversity conservation, further promoted by employing logistic support for education, research, training, and monitoring, becomes the basis for sustainable development – the economic development reconciled with the environment.

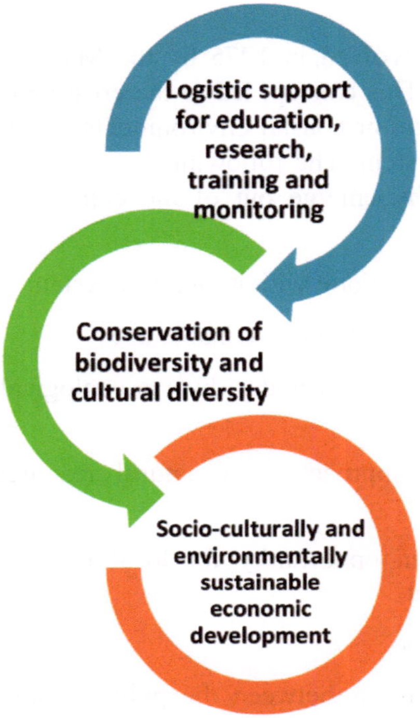

Fig. 6.4: Planning and management of a biosphere reserve

A biosphere reserve maintains these functions based on the three zones it is demarcated in. These zones are the core zone, the buffer zone, and the transition zone as depicted in Fig. 6.5. It is the core area which is the core conservation area functioning to conserve biodiversity at the landscape, ecosystem, species, and genetic levels. Human intervention in the core area is strictly prohibited. Natural evolutionary processes in this zone prevail strongly and the zone is thought to be ecologically healthy, vibrant, and sustainable. However, interesting scientific research in this zone may be carried out.

The buffer zone surrounds the core zone and is itself surrounded by the transition zone. This zone is sandwiched between the zone meant for conservation and the zone with human activities. This zone is often used for education, scientific research, training, monitoring, etc. The zone is also open for tourism.

The outermost area of a biosphere reserve comprises what is meant for human intervention for fostering developmental activities for economic gains. The developmental activities encouraged are the sustainable ones, that is, ecologically sound, environmentally safe, and socio-culturally just.

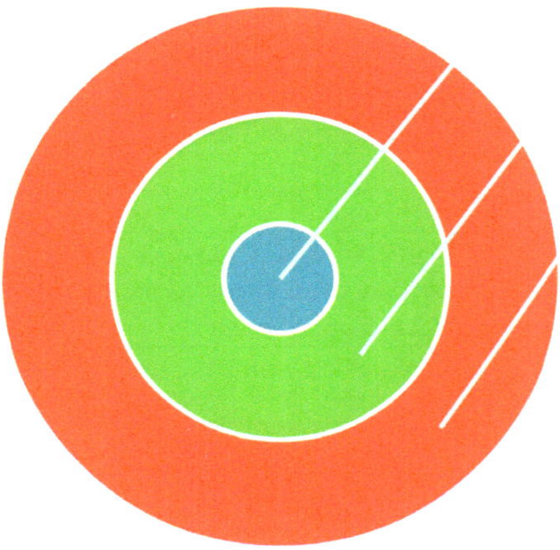

Fig. 6.5: The three distinguishable but integrated zones of a biosphere reserve

A list of some important biosphere reserves in India is presented in Table 6.6. Many of the biosphere reserves in India are protected by UNESCO. Five of the protected areas in India have been recognized as UNESCO Heritage Sites (Table 6.7.)

Table 6.6: Important biosphere reserves in India

Name and location	Area in km^2	Important animals
Great Rann of Kutch (Gujarat)	12454	Indian wild ass
Gulf of Mannar (Tamil Nadu)	10500	Dugong
Sunderban Biosphere Reserve (West Bengal)	9630	Royal Bengal Tiger
Cold desert (Himachal Pradesh)	7770	Snow leopard
Nanda Devi Biosphere Reserve (Uttarakhand)	5860	Snow leopard, Himalayan black bear
Nilgiri Biosphere Reserve (Tamil Nadu, Kerela, Karnataka)	5520	lion-tailed macaque,tiger,Nilgiri Tahr
Dihang Dibang (Arunachal Pradesh)	5112	Mishmi takin, musk deer
Seshachalam Hills (Andhra Pradesh)	4755	Slender loris
Simplipal Biosphere Reserves (Odisha)	4374	Gaur, royal Bengal tiger, Asian elephant
Pachmarhi Biosphere Reserve (Madhya Pradesh)	4189.78	Giant squirrel, flying squirrel
Achanakmar-Amarkantak Biosphere Reserve (Madhya Pradesh, Chattisgarh)	3835	Four-horned antelope, Indian wild dog, sarus crane, white-rumped vulture, sacred grove bush frog

Name and location	Area in km²	Important animals
Agasthyamalai Biosphere Reserve (Kerela ,Tamil Nadu)	3500	Nilgiri tahr, Asian elephant
Manas (Assam)	2837	Asiatic elephant, tiger, Assam roofed turtle, hispid hare, golden langur, pygmy hog
Khangchendzonga (Sikkim)	2620	Snow leopard, Red Panda
Great Nicobar (Andaman and Nicobar Islands)	885	Saltwater crocodile
Nokrek (Meghalaya)	820	Red Panda
Dibru-Saikhowa(Assam)	765	White-winged wood duck, water buffalo, black-breasted parrot bill, tiger, capped langur
Panna (Madhya Pradesh)	298.98	Tiger, chital, chinkara, sambhar and sloth bear

Table 6.7: UNESCO Heritage Sites in India

S.No.	Protected Area (UNESCO Heritage Site)	Location (State of India)	Area in km²
1.	Kaziranga National Park	Assam	430
2.	Keoladeo Ghana National Park	Rajasthan	29
3.	Manas Wildlife Sanctuary	Assam	950
4.	Sundarbans National Park	West Bengal	2585
5.	Khangchendzonga National Park	Sikkim	1784

Sacred Forests

Sacred forests or sacred groves represent an alternative approach to conservation that involves the principles emanating from a diversity of prevailing cultures or social taboos rather than the approaches based on the legal framework. Protection and conservation of biodiversity in the environments of sacred forests have been a tradition in many Asian countries. India is especially known for this tradition. Sacred seeds, sacred trees, sacred forests, sacred rivers, sacred lakes, etc. are the various notions underlining an outstanding life-caring system in Indian cultural practices. Almost every temple in India has a sacred tree nearby that is never pruned and is used as a breeder tree. Its fruits and seeds are dispersed to other places for multiplication.

The Hindus uphold religious sanctity for some species-specific forests and protect them for conservation, performing religious rituals on particular occasions. Human intervention in the sacred forests is minimum. Logging and poaching in these community-protected forests are strictly prohibited. Thus, there are the least chances of ecological degradation in these forests. As a result, such community-protected forests, in a sense, represent pristine forests. These forests are healthy and vibrant and serve as home to native biodiversity.

Many sacred forests are often surrounded by degraded landscapes. In that case, the sacred forests serve as oases of greenery with native species flourishing in the natural environment.

India embraces enormous cultural diversity as well as sacred forests being managed by diverse cultures. The number of sacred forests in India might be well over 100,000, according to a study by Ormsby et al. (2010). This approach of in-situ conservation has greater acceptance among local people. Many states of India, notably Himachal Pradesh, Madhya Pradesh, Bihar, Maharashtra, Manipur, Meghalaya, Kerala, Karnataka, and Tamil Nadu, have many sacred forests being managed by local communities. Many rare, endangered, and endemic species are prospering in these people-protected ecosystems.

The sacred forests in India are known by various names depending upon geographical and cultural zone. For example, in the states of Madhya Pradesh and Himachal Pradesh, they are referred to as Dev Van (God's forest). In other Indian states, the community-protected sacred forests are locally known as *Sarnas* in Bihar, *Oran* in Rajasthan, *Devarakadu* in Karnataka, *Kavu* in Kerala, *Devrai* or *Dev Van* in Maharashtra, *Lai Umang* in Manipur, *Law Kyntang* or *Law Lyngdoh* in Meghalaya, and *Sapra Kavu* in Tamil Nadu. Cultural rituals may be different in various states or geographical areas, but the ways and means the sacred forests are protected are almost the same. Again, the species considered sacred might also vary from place to place, but that is crucial for biodiversity conservation.

Celebrating biodiversity is a way of life in Hinduism. There are several festivals throughout the year which are devoted to particular species of plants. Every Hindu festival is associated with a plant species and/ or foods derivable from a variety of plants. *Azadirachta indica*, *Ficus bengalensis*, *Ficus religiosa*, *Aegle marmelos*, *Musa paradisiaca* etc. are the main trees celebrated on different occasions in India (Anubhav et al. 2009). A glimpse of these celebrations is shown in Fig. 6.6. Every devout Hindu family has Holy Basil popularly known as Tulsi (*Ocimum sanctum* and *Ocimum tenuiflorum*) planted at home. This herbaceous species is worshipped every day in a Hindu family. There are numerous other plants – trees, shrubs, and herbs – which are given due protection as they are sources of many life-saving medicines. Pseudo cereals, amaranth, and buckwheat, are commonly used for eating on the occasion of observing a fast.

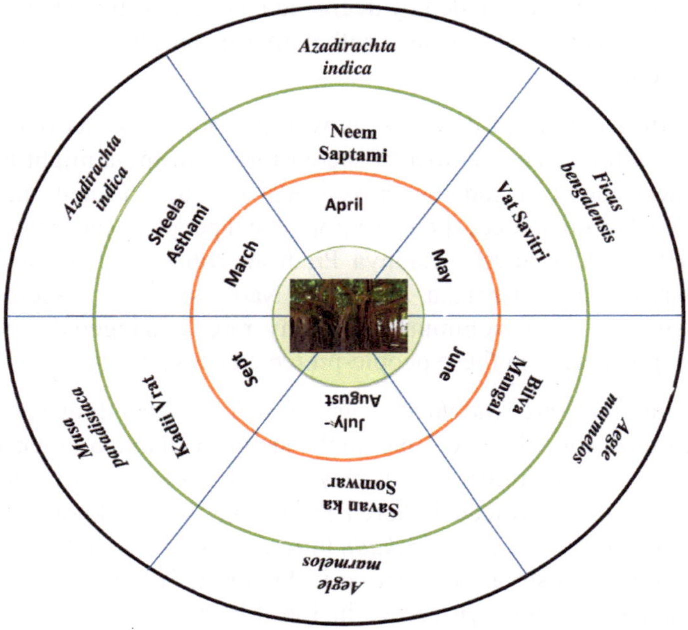

Fig. 6.6: Religious festivals associated with sacred trees

Sacred groves are not only sacred ecosystems functioning as a rich repository of nature's unique biodiversity but also a product of the socioecological philosophy our forefathers have been cherishing since the olden days. These ecosystems do not just help conserve valuable biodiversity, soil, and water, but are also critical in regulating weather and climate cycles so vital for life to blossom and flourish on the planet. This kind of informal protection takes care of the protection of rare, endangered, and vulnerable species in their natural habitats that are often missed from protection through formal protection measures.

Sacred Lakes

Many water bodies in India are also protected by communities. A glaring example is that of Khecheopalri Lake in Sikkim, which was declared a sacred lake by the people of the area. This sacred lake is taken care of by the local community for the protection of the local flora and fauna of the local aquatic ecosystems.

Bindu Sarovar and Narayan Sarovar in Gujarat, Pampa Sarovar in Karnataka, Pushkar Lake in Rajasthan, and Manas Sarovar or Mansarovar (now in Tibet) are some of the most sacred lakes in India. Tibet is also known for its sacred lakes. The main sacred lakes of Tibet are Lake Namtso (also called the Heavenly Lake), and Basum Lake. Tangra Yumco, Lake Rakas Tal (also called the Devil Lake), and Lamu Nacuo (also called the Goddess Skull Lake).

It is due to the notion of sacredness that the sacred lakes are managed in such a way that they are kept pollution free and serve as healthy habitats for aquatic life. Sacred groves and sacred lakes both significantly contribute to protecting and conserving biodiversity with overwhelming participation by the concerned communities.

Ex-Situ Conservation

The human species, as we noted earlier, has emerged as a custodian of nature's biodiversity. Ever increasing and intensifying human intervention into natural ecosystems, conservation of all the species and sub-species has become difficult in their natural habitats. Therefore, conservation of biodiversity at places away from their natural habitats, or ex situ conservation, has become of critical importance. The establishment of germplasm banks or gene banks is the main strategy of ex situ conservation. The gene banks include botanical gardens, zoos, genetic resource centers, seedlings, seed, tissue culture, and DNA banks.

Gene Bank

A gene bank is a biorepository where seeds and cuttings or vegetative parts from a large variety of plants are stored. The main role of plant gene banks is to preserve genetic diversity in the form of seeds, cuttings, sperms, and eggs. The genetic material in a gene bank is preserved in a variety of ways, such as i) by creating artificial ecosystems, ii) in controlled conditions, iii) in appropriate nutrient media, and iii) by freezing at very low temperatures, up to –196° Celsius, in liquid nitrogen.

Preservation of animal germplasm is done by the freezing of sperms and eggs in zoological freezers. For corals, fragments are taken and stored in water tanks under controlled conditions. The genetic material preserved in a gene bank is meant to be made available along with associated information for future use in research and plant and animal breeding.

Cryopreservation

The cryopreservation (Greek: krúos = icy cold or frost) is the process of storage or preservation of biological samples in extremely cold or sub-zero temperatures, at −196°C, using liquid nitrogen. Temperatures up to this level help maintain the survivability of the biological material with the thawing of physiological conditions. Cooling at extremely low temperatures causes the formation of ice crystals inside cells leading to rupture of the cell membrane and cell death. Cryopreservation prevents the formation of ice crystals inside cells avoiding cell death. Biological material including organelles, cells, tissues, extracellular matrix, and organs are preserved and conserved using cryopreservation.

Botanical Gardens

A botanical garden, or botanic garden, is a place dedicated to biodiversity conservation with the collection, cultivation, preservation, and display of a wide range of plants. This is a garden, especially of living plants designed chiefly to illustrate relationships within plant groups for scientific study, research, and ex situ conservation of plant biodiversity. The collection includes specialist plants, for example, cacti and other succulent plants, ornamental herbs and ornamental shrubs, or plants from a particular geographical region of the world, etc.

A botanical garden involving only trees is referred to as an arboretum. An arboretum is often associated with a zoo or wildlife. The botanical gardens (and arboreta) are often managed by scientific research institutes or universities. In some cases, they are independent institutes in themselves and have their agenda of research, education, and conservation. There are more than 2000 botanical gardens and arboreta in the world with more than 80,000 species of plants under ex situ conservation. Many botanic gardens have their seed banks, tissue culture facilities as well as technologies worth promoting ex situ conservation. Some prominent botanic gardens in the world include:

London Kew Garden, UK

Longwood Gardens, USA

Jardin Botanique Montreal, Canada

Hawaii Tropical Botanical Garden

Orto Botanico di Padova, Italy

Botanischer Garten München

Kirstenbosch Botanical Garden, South Africa

Singapore Botanical Gardens

Sydney Royal Botanical Garden, Australia

New York Botanical Garden

Established in 1544, the Orto Botanico di Pisa has the credit of being the world's oldest garden and was operated by the University in Pisa in Italy. Founded in 1545 and run by the University of Padua, the Orto Botanico di Padova in Italy is sometimes regarded to be the oldest academic botanic garden because it was not moved from its original place. The credit for the world's botanic garden with the largest collection of living plants, numbering about 50000, goes to the London Kew Garden. In Asia, the largest botanic garden is the Gyeongsangbuk-do Arboretum. Founded in the year 2001, spread over some 3,222 ha area and comprising 179,226 plants belonging to 1,510 different species, this botanic garden is located in South Korea.

The history of botanical gardens in India begins with Indian Botanical Garden, or Royal Botanic Garden or Calcutta Botanic Garden, which was established in Shibpur, Kolkata in 1786. Spread over 109 hectares (270 acres), the Indian Botanical Garden, later named Acharya Jagdish Chandra Bose Indian Botanic Garden, has also the credit to be the largest one in India. The world's largest banyan tree is the main attraction of this botanic garden.

The Zoos

Establishment of the zoos is the common strategy supporting the ex situ conservation of wild animals. The zoos are situated away from the wilderness, often within or adjacent to human habitats, and are also open to visitors and tourists. Varieties of animal species have been protected in these human-managed ecosystems. The zoos are also used as places for recreation. However, wild animals in zoos are often reported to be under stress and this ex situ conservation method does not provide an environment conducive to the natural protection, conservation, and growth of wild animal protection. Captive breeding programs run in professionally managed zoos often do not go successful due to the lack of natural environment the animals need for the development of their natural traits and behavior.

There are some 800 professionally managed zoos of appreciable size in the world, giving conservation cover to more than 3000 species of mammals, birds, reptiles, amphibians, etc. Many of the animal species being conserved in the zoos include the rare and endangered species, and also those threatened by extinction or those on the verge of extinction.

The oldest zoo in the world is the Vienna Zoo, Tiergarten Schonbrunn, meaning Animal Garden. Founded in 1752, the Vienna Zoo continues to be operating successfully to date. The other important and often-talked about zoos conserving a variety of fauna are the National Zoological Gardens of Africa, London Zoo, San Diego Zoo, Moscow Zoo, Toronto Zoo, Columbus

Zoo and Aquarium, The Bronx Zoo, Berlin Zoological Garden, Dallas Zoo, Philadelphia Zoo, Oakland Zoo, Detroit Zoo, Los Angeles Zoo, etc.

Hotspots of Biodiversity

Earth is the home of biodiversity, perhaps the only home in the cosmos known so far. The distribution of biodiversity on the planet, however, is not uniform. It is obvious as biodiversity is an attribute of a complex of environmental factors. Certain regions of the Earth are witnessed to harbor a very large number of species. The regions prospering with a very high measure of biodiversity are referred to as megadiversity zones. Some twelve countries with 60–70 percent of the world's biodiversity have been identified as megadiversity countries. India is one of the megadiversity countries contributing some eight percent of living species to global biodiversity despite occupying only 2.4 percent of the world's geographical area.

Norman Myers (1934–2019) elucidated the concept of hotspots to designate priority areas for in situ conservation and then elaborated the same in 1990 and 2000 (Myers 1988, 1990; Myers et al. 2000). The hotspots of biodiversity are defined as the richest and the most threatened reservoirs of plant and animal life on Earth. There are two major criteria based on which a hotspot of biodiversity is determined. These are:

1. Number of endemic species, that is, the species that are not found in other areas: a hotspot of biodiversity must contain at least 1,500 species of vascular plants (> 0.05% of the world's total) as endemics;
2. The degree of threat is measured in terms of habitat loss: a hotspot should have lost at least 70% of its original natural vegetation.

There are some three dozen biodiversity hotspots identified the world over that strictly fulfill the above-mentioned criteria, with the purpose of in situ biodiversity conservation. These hotspots represent just 2.3% of the Earth's land surface but support approximately 60% of the Earth's plant species and 42% of all terrestrial vertebrate species—mammals, birds, reptiles, and amphibian species.

Many of the biodiversity hotspots exceed the two criteria mentioned in the definition. For example, Tropical Andes in South America and Sundaland in Southeast Asia, the two biodiversity hotspots, have approximately 15,000 endemic plant species. Vegetation loss in some of the hotspots has exceeded to an extent of a whopping 95%.

Megadiverse Countries

There are 17 countries identified the world over to be credited as megadiverse countries. These are:

1. Brazil
2. Columbia
3. Ecuador
4. Mexico
5. Peru
6. Venezuela
7. United States
8. The Congo
9. Madagascar
10. South Africa
11. India
12. China
13. Malaysia
14. Philippines
15. Indonesia
16. Papua New Guinea
17. Australia

The biodiverse countries represent more than two-thirds of all the known life forms. Most tropical rainforests and coral reefs are also included in these megadiverse countries. The megadiverse countries possess less than 10% of the global surface but support more than 70% of the global biodiversity. The biodiversity hotspots falling in various biogeographic regions are presented in Table 6.8:

Table 6.8: Biodiversity hotspots of the world

Biogeographic Region	Biodiversity hotspot
South America	Tropical Andes Atlantic Forest Tumbes-Chocó-Magdalena Cerrado Chilean Winter Rainfall-Valdivian Forests
The Caribbean	Caribbean Islands
North and Central America	Mesoamerica California Floristic Province Madrean Pine-Oak Woodlands North American Coastal Plain
Africa	Madagascar and the Indian Ocean Islands Coastal Forests of Eastern Africa Guinean Forests of West Africa Cape Floristic Region Succulent Karoo Maputaland-Pondoland-Albany Eastern Afromontane Horn of Africa
Europe	Mediterranean Basin
West Asia	Caucasus Irano-Anatolian
East Asia	Mountains of Southwest China Japan
South Asia	Indo-Burma, India and Myanmar Western Ghats and Sri Lanka Eastern Himalayas
Central Asia	Mountains of Central Asia
South East Asia and Asia Pacific	Sundaland and Nicobar Islands of India Wallacea Philippines Southwest Australia New Caledonia New Zealand Polynesia-Micronesia East Melanesian Islands Eastern Australian Temperate Forests

Source: Mittermeier, R. A., Myers, N., Gil, P. R. and Mittermeier, C. G. 1999. *Hotspots: Earth's Biologically Richest and Most Endangered Terrestrial Ecoregions*. Monterrey, Mexico: Cemex, Conservation International and Agrupacion Sierra Madre.

Just one year after Norman Myers published his work on the biodiversity of hotspots, Conservation International, an environmental organization, adopted the idea of protecting the biogeographical regions exceptionally rich with biodiversity but threatened by various factors of destruction.

Five of the global biodiversity hotspots are located in India. These hot spots reveal a very high degree of endemism and inhabit a large number of diverse flowering plants, mammals, reptiles, amphibians, and swallow-tailed butterflies. As the geo-biological regions know no international boundaries, the biodiversity hotspots in India extend to the boundaries of many neighboring countries and those located in the Indian Ocean.

- **Himalayas:** The entire Indian Himalayan region extending up to the parts of Pakistan, Tibet, Nepal, Bhutan, and Myanmar
- **Indo-Burma:** Entire Northeastern India (except Assam and Andaman Islands) extended to Myanmar, Thailand, Vietnam, Cambodia, Laos, and southern China
- **Eastern Himalayas:** Part of the two hotspots, viz. Indo-Burma and newly distinguished Himalayas; includes North-East India extended to Bhutan and southern, central, and eastern Nepal; originally part of the Indo-Burma biodiversity hotspot; Numerous primitive angiosperm families, such as *Magnoliaceae* and *Winteraceae*, and primitive genera of plants, e.g., *Betula* and *Magnolia*
- **Western Ghats:** Entire Western Ghats parallel to the Western Coast of the Indian Peninsula for about 1600 km covering parts of the Indian states of Karnataka, Maharashtra, Kerala, and Tamil Nadu with two main biodiversity centers, viz. i) the Agastryamalai hills and Silent Valley, and ii) the Amambalam Reserve; extended to Sri Lanka
- **Sundaland**: Nicobar Group of Islands extended to Malaysia, Singapore, Philippines, Indonesia, Brunei

International Efforts for Biodiversity Conservation

The Earth Summit held in June 1992 in Rio de Janeiro has been a milestone in the history of environmental movements that began after the publication of Rachel Carson's *Silent Spring* in 1962 and the first-ever conference on Environment and Development in Stockholm in 1972. The concept of Sustainable Development laid down in the Brundtland Commission Report *Our Common Future* in 1987 called for reconciliation of environment and development. The Earth Summit was very special in the sense that it was the first to raise biodiversity-related issues on an international platform. The Biodiversity Convention resulted from the Earth Summit came into force on 29 December 1993. The Biodiversity Convention had three key objectives, namely:

1. Conservation of biodiversity
2. Sustainable use of biodiversity
3. Fair and equitable sharing of benefits arising out of the utilization of genetic resources

Several projects focusing on the conservation of biodiversity and appropriate measures for the development of biosphere reserves, national parks, and wildlife sanctuaries are being supported by international organizations, like the International Union for the Conservation of Nature (IUCN) and World Wildlife Fund for Nature (WWF). Projects relating to ex-situ and on-farm biodiversity conservation in agriculture, forestry, and livestock sectors are supported by the Food and Agriculture Organization (FAO). Numerous non-government organizations all over the world are also engaged in biodiversity protection, conservation, and amelioration.

Several international organizations are engaged in various ways to conserve biodiversity hotspots. However, only a small proportion of the land area falling within the defined biodiversity hotspots is currently protected. Some of the organizations committed to protecting and conserving biodiversity hotspots and biodiversity within these hotspots are:

The National Geographic Society

Critical Ecosystem Partnership Fund (CEPF)

World Wide Fund for Nature (Global 200 Ecoregions)

Plant Life International

Birdlife International

Alliance for Zero Extinction

Summary

There are two major challenges in implementing conservation strategies: limited space and human pressure on natural resources. Two basic strategies of biodiversity conservation applied in the world are in-situ conservation and ex-situ conservation. Conservation of the biological diversity in its natural habitats is the in-situ conservation strategy. Protected areas include earmarked and defined geographical areas specially dedicated to the protection and maintenance of biodiversity in natural environs or wilderness. The protected areas include national parks and wildlife sanctuaries and are protected and managed by employing legal measures and/ or by other effective means. There are as many as 202,467 protected areas throughout the world as of October

2017 according to the World Database on Protected Areas (WDPA) which cover about 200,000,000 km^2 area, or 14.7% of the world's land, Antarctica excluded. The Convention on Biodiversity under the Aichi Biodiversity Targets has set 17% area to be covered for biodiversity conservation in 2020. There is still a gap of 2.3% to achieve the target of protected areas. Apart from the land area, some 10% of the territorial waters are also protected. A protected area under the in-situ conservation strategy is ecologically more complex, healthier, more vibrant, more resilient, more productive, and more sustainable than an anthropogenic one managed under an ex-situ conservation strategy.

A national park is meant for the welfare of wildlife. Human activities like agriculture, forestry, grazing, or land use change are not allowed. National parks numbering 6,555 throughout the world meet the criterion defined by IUCN. A sanctuary is meant to be an area legally protected or reserved only for the conservation of wild animals. Habitats of wild animals and their surroundings are given due protection for their safety, comfortable living, free movement, and breeding. Natural Park of the Coral Sea encompassing an area of 129,49,94 km^2 and Pacific Remote Islands Marine National Monument with an area of 1,269,094 km^2 are among the world's largest animal wildlife sanctuaries. The concept of a biosphere reserve was evolved in 1975 by the Man and Biosphere (MAB) program of UNESCO. This concept led to the creation of a special category of protected areas – land or coastal environments – with people as an integral component of the system. At present, there are 701 biosphere reserves in the world within the boundaries of 124 countries. A biosphere reserve maintains these functions based on the three zones it is demarcated into. These zones are the core zone, the buffer zone, and the transition zone.

Sacred forests or sacred groves represent an alternative approach to conservation that involves the principles emanating from a diversity of prevailing cultures or social taboos rather than the approaches based on the legal framework. Protection and conservation of biodiversity in the environments of sacred forests have been a tradition in many Asian countries. India is especially known for this tradition. Celebrating biodiversity is a way of life in Hinduism. There are several festivals throughout the year which are devoted to particular species of plants. Every Hindu festival is associated with a plant species and/or foods derivable from a variety of plants. Many water bodies in India are also protected by communities.

Conservation of biodiversity at places away from their natural habitats, or ex situ conservation, has become of critical importance. The establishment of germplasm banks or gene banks is the main strategy of ex situ conservation.

The gene banks include botanical gardens, zoos, genetic resource centers, seedlings, seed, tissue culture, and DNA banks. Cryopreservation or cryo conservation is the process of storage or preservation of biological samples in extremely cold or sub-zero temperatures, at −196°C, using liquid nitrogen.

The hotspots of biodiversity are defined as the richest and the most threatened reservoirs of plant and animal life on Earth. There are some three dozen biodiversity hotspots identified the world over that strictly fulfill the above-mentioned criteria, with the purpose of in situ biodiversity conservation. These hotspots represent just 2.3% of the Earth's land surface but support approximately 60% of the Earth's plant species and 42% of all terrestrial vertebrate species—mammals, birds, reptiles, and amphibian species.

There are 17 megadiverse countries identified in the world. The biodiverse countries represent more than two-thirds of all the known life forms. Most tropical rainforests and coral reefs are also included in these megadiverse countries. The megadiverse countries possess less than 10% of the global surface but support more than 70% of the global biodiversity.

The Earth Summit was very special in the sense that it was the first to raise biodiversity-related issues on an international platform. The Biodiversity Convention resulted from the Earth Summit came into force on 29 December 1993. It had three key objectives, namely: conservation of biodiversity, sustainable use of biodiversity, and fair and equitable sharing of benefits arising out of the utilization of genetic resources.

Several projects focusing on the conservation of biodiversity and appropriate measures for the development of biosphere reserves, national parks, and wildlife sanctuaries are being supported by international organizations, like IUCN and WWF. Projects relating to ex-situ and on-farm biodiversity conservation in agriculture, forestry, and livestock sectors are supported by FAO. Numerous non-government organizations all over the world are also engaged in biodiversity protection, conservation, and amelioration.

Exercises

A. Multiple Choice Questions

1. *In-situ* conservation refers to the conservation of biodiversity in

    a. Its natural habitats   b. Cultivated lands

    c. Botanical gardens   d. Home gardens

2. The ex-situ conservation of genetic resources involves

a. Contour plowing
b. Shifting cultivation
c. Minor irrigation schemes
d. None of the above

3. How much area should be covered for biodiversity conservation, according to the Convention on Biodiversity under the Aichi Biodiversity Targets in 2020?

a. 10%
b. 14%
c. 17%
d. 22%

4. Which of the following countries has a minimum protected area?

a. Australia
b. New Zealand
c. Afghanistan
d. Laos

5. Which of the following countries has the largest protected area for biodiversity conservation?

a. New Caledonia
b. India
c. Bhutan
d. Poland

6. Which is the oldest protected area in the world?

a. The Tobago Main Ridge Forest Reserve
b. The Yellowstone National Park
c. Royal National Park, Australia
d. Banff National Park, Canada

7. The Sequoia National Park is located in

a. Australia
b. New Zealand
c. USA
d. Russia

8. The country with most of the National Parks, as many as 285, is

a. Australia
b. USA
c. Russia
d. Denmark

9. Northeast Greenland National Park in Denmark has the reputation of being the

a. Smallest in size
b. Largest in size
c. Oldest of the world
d. Driest desert

10. Jim Corbett National Park is

a. The largest in India
b. The oldest in India
c. The smallest in size in India
d. Famous for rhinoceros

11. Match the terms of column A with appropriate terms of Column B and select the most appropriate answer:

Column A	Column B
A. Kaziranga National Park	i. Kamrup (Assam)
B. Desert National Park	ii. UNESCO
C. Manas Wildlife Sanctuary	iii. Biosphere Reserve
D. Agasthyamalai	iv. Rhinoceros
E. MAB	v. Great Indian Bustard
F. – 196°C	vi. Cryopreservation

a. A-vi, B-i, C-v, D-ii, E-iv, F-iii
b. A-i, B-vi, C-ii, D-v, E-iii, F-iv
c. A-iv, B-v, C-i, D-iii, E-ii, F-vi
d. A-vi, B-v, C-iv, D-iii, E-ii, F-i

12. Which of the following is the matching one?

a. Sundarban – Rhino
b. Gir – Lion
c. Keoladev – Elephant
d. Kaziranga – Musk deer

13. Match the terms of column A with appropriate terms of Column B and select the most appropriate answer:

Column A	Column B
A. Chiru	i. William Goud
B. Biodiversity Hot Spots	ii. Exotic species
C. Green Revolution	iii. Anti-cancer drug
D. Yew tree	iv. Shahtoosh
E. Eichhornia	v. Norman Myers
F. Project Tiger	vi. 1973

a. A-vi, B-i, C-v, D-ii, E-iv, F-iii
b. A-iv, B-v, C-i, D-iii, E-ii, F-vi
c. A-i, B-vi, C-ii, D-v, E-iii, F-iv
d. A-vi, B-v, C-iv, D-iii, E-ii, F-i

14. Which is the world's largest animal wildlife sanctuary?
    a. Pacific Remote Islands Marine National Monument
    b. Natural Park of the Coral Sea
    c. Kavango-Zambezi Transfrontier Conservation Area
    d. Papahānaumokuākea Marine National Monument
15. In 1975, the Man and Biosphere (MAB) Programme of UNESCO evolved the concept of
    a. National Park
    b. Wildlife Sanctuary
    c. Project Tiger
    d. Biosphere Reserve
16. Which zone of a biosphere reserve is often used for education, scientific research, training, monitoring, etc., and is also open for tourism?
    a. Core zone
    b. Buffer zone
    c. Transition zone
    d. None of the zones
17. The credit of the world's botanic garden with the largest collection of living plants, numbering about 50000, goes to the
    a. London Kew Garden, UK
    b. Acharya Jagdish Chandra Bose Indian Botanic Garden, India
    c. Kirstenbosch Botanical Garden, South Africa
    d. Sydney Royal Botanical Garden, Australia
18. The world's largest banyan tree is the main attraction of this botanic garden. This is
    a. Acharya Jagdish Chandra Bose Indian Botanic Garden, India
    b. Botanical Garden of India Republic, India
    c. Singapore Botanical Gardens, Singapore
    d. Orto Botanico di Padova, Italy
19. Who did elucidate the concept of biodiversity hotspots?
    a. E. P. Odum
    b. R. Lindeman
    c. Charles Elton
    d. Norman Myers

20. Which among the following is not a megadiverse country?

a. India
b. Brazil
c. Ecuador
d. Italy

Answers

1. **a**, 2. **d**, 3. **c**, 4. **c**, 5. **a**, 6. **a**, 7. **c**, 8. **a**, 9. **b**, 10. **b**, 11. **c**, 12. **b**, 13. **b**, 14. **b**, 15. **d**, 16. **b**, 17. **a**, 18. **a**, 19. **d**, 20. **d**

B. Short-Answer Questions

1. Differentiate between in-situ and ex-situ conservation of biodiversity.
2. What is a protected area?
3. Name five oldest national parks in the world.
4. Define different zones in a biosphere reserve.
5. Name five biosphere reserves in India.
6. What is cryopreservation?
7. What do you understand by an arboretum?
8. List the names of some famous botanical gardens of the world, mentioning their locations.
9. Name the megadiverse countries of the world.
10. Name the hotspots of biodiversity in India.

C. Long-Answer Questions

1. Explain the biodiversity conservation measures that are adopted at all levels.
2. What would be the phenomenal ecological impacts of biodiversity conservation?
3. What are the main objectives and roles of a biosphere reserve?
4. Explain the international efforts for biodiversity conservation.
5. Write short notes on the following:
   a. Sacred Forests and Sacred Lakes
   b. Botanical Garden
   c. Biodiversity Hotspots

References

Anubhav, Singh, K., Rastogi, A. and Singh, Vir. 2009. Sacred groves in India: celebrating sanctity of life through biodiversity conservation. In: Gautam, P. L., Singh, Vir and Melkania, U. (eds.) *Ecosystem Diversity and Carbon Sequestration: Climate Change Challenges and a Way Out for Ushering in a Sustainable Future*. New Delhi: Daya Publishing House. pp. 89-101.

Mittermeier, R. A., Myers, N., Gil, P. R. and Mittermeier, C. G. 1999. *Hotspots: Earth's Biologically Richest and Most Endangered Terrestrial Ecoregions*. Monterrey, Mexico: Cemex, Conservation International and Agrupacion Sierra Madre.

Myers, N. 1988. Threatened biotas: "Hot spots" in tropical forests. Environmentalist, 8: 187-208. doi:10.1007/BF02240252

Myers, N. 1990. The biodiversity challenge: expanded hotspots analysis. *Environmentalist*, 10: 243–256.

Myers, N., Mittermeier, R. A., Mittermeier, C. G., da Fonseca, G. A. B. and Kent, J. 2000. Biodiversity hotspots for conservation priorities. *Nature* 403: 853–858. https://doi.org/10.1038/35002501

Ormsby, A. A. and Bhagwat, S. A. 2010. Sacred forests of India: a strong tradition of community-based natural resource management. *Environmental Conservation*, 37(3): 320-326. doi:10.1017/S0376892910000561

Singh, Vir. 2019. *Fertilizing the Universe: A New Chapter of Unfolding Evolution*. London: Cambridge Scholars Publishing. 285pp.

Web Resources

http://www.wiienvis.nic.in/Database/Protected_Area_854.aspx

https://synopsisias.in/ene-12-biodiversity/

https://www.biodiversitya-z.org/content/iucn-protected-area-management-categories.pdf

https://www.easy-grow.co.uk/10-amazing-botanical-gardens-in-the-world/

https://www.environment.gov.au/biodiversity/conservation/hotspots

https://www.indexmundi.com/facts/indicators/ER.LND.PTLD.ZS/rankings

7

Biodiversity and Sustainability

Sustainability does not denote a fixed state of things. It is a dynamic phenomenon: a phenomenon that tends to advance towards a higher state of being. The range of a journey of sustainability extends from nothing to everything, from a simple system to a more complex system, or from an unstable or less stable system to a stable or more stable system. Speaking in ecological terms, the transformation of unproductive desert land into a forest in ecological climax over time is how the phenomenon of sustainability works.

The term 'sustainable development' was first of all used in 1980 in the World Conservation Strategy (IUCN-UNEP-WWF 1980). However, this term was brought into the focus of the world for the first time in the United Nations World Commission on Environment and Development (WCED) Report, popularly known as the Brundtland Commission Report *Our Common Future* in 1987. Sustainable Development in the WCED Report has been defined as *Development that meets the needs of the present without compromising the ability of the future generations to meet their own needs.*

Sustainable development, thus, looks at the state of socioeconomic well-being through environmental issues. Sustainability, sometimes used as synonymous with sustainable development, generally refers to the human ability to live constantly in tune with the biosphere's capacities and biodiversity is the crux for building up the biosphere's capacities.

Ecological Sustainability: A Vital Dimension of Sustainability

The concept of sustainability is often distorted in favor of the actors involved. For example, it is delinked from its deeper meaning spelling out its ecological dimensions and is often targeted for economic benefits. Of course, sustainability may have many dimensions including, for example, socioeconomic sustainability. However, no dimension other than the ecological dimension is independent. All dimensions, eventually, are expressed through the ecological dimension. As Singh (2019, 2020) logically and categorically puts it, there can be no sustainability without ecological sustainability of its own.

A sustainable economy cannot be ascertained without ecosystem considerations. All sorts of livelihood and economic systems ultimately survive on and function and sustain through ecosystem functioning. A vibrant and sustainable economy can be ensured only through a healthy and vibrant ecosystem, and a healthy and vibrant ecosystem depends on the biodiversity it provides refuge to, nourishes, and sustains. Biodiversity, on the other hand, structures and maintains an ecosystem. Biodiversity, in essence, is the basis of environmental sustainability which, in turn, serves as the basis of other dimensions of sustainability.

Biodiversity-Sustainability Links

Biodiversity builds up, induces, enhances, supports, and maintains sustainability. This is nature's law. Biodiversity enhances nature's resilience, that is, nature's capacity to recuperate from physical and biological shocks (Singh 2019, 2020). The higher the level of biodiversity in nature the higher the degree of resilience. In natural ecosystems, sustainability is attained via resilience. The higher the degree of resilience the higher the state of sustainability.

Biodiversity provides a wide range of choices for the predators – herbivores as well as carnivores – to feed on and survive in an ecosystem. For example, different insects prosper on different types of plants. If there is a very high level of plant diversity in an ecosystem, they would all survive and flourish without any considerable pressure of defoliation by the insects. Monocultures are often destroyed by species-specific pests.

Monocultures introduce vulnerability in ecosystems. These are liable to be damaged by physical factors like strong winds, storms, heavy rains, fires, etc., and by biotic pressures, such as intensive grazing, insects, plant pathogens, etc. Heterogeneous ecosystems, contrarily, are invariably more resilient and, consequently, more sustainable.

Productivity is also regarded as one of the indicators of sustainability. From this point of view, an ecosystem harboring an extreme degree of biodiversity would be sustainably more productive than the monocultures.

Biodiversity attributes translating into the ecosystem and socioeconomic sustainability are complementary to each other; one influences and is influenced by the others (Fig. 7.1). Based on the natural links between biodiversity and sustainability as discussed above, ecologically sustainable and vulnerable ecosystems can be distinguishable. The ecosystems featuring a higher level of biodiversity are more sustainable than the ones with lower levels of biodiversity. In other words, the ecosystems with a lesser level of biodiversity are ecologically more vulnerable than the ones with a higher level of biodiversity (Table 7.1).

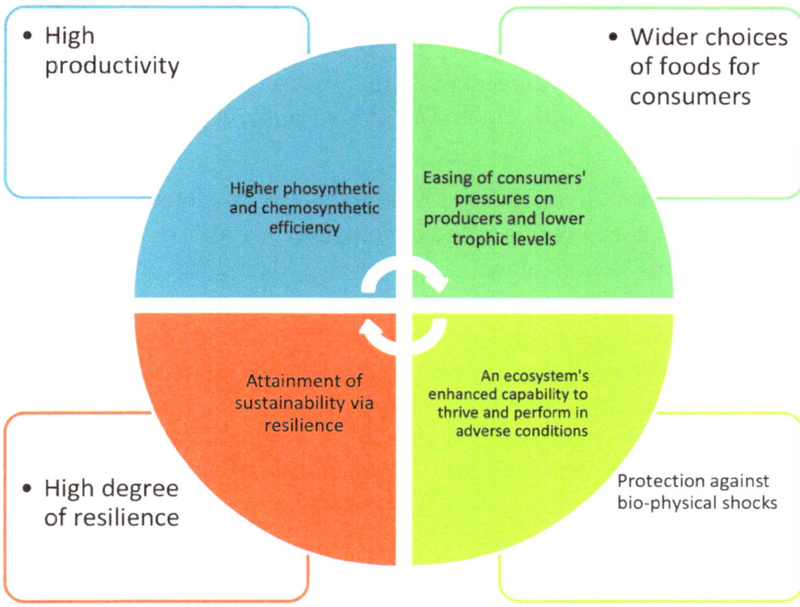

Fig. 7.1: Biodiversity-sustainability links: Biodiversity attributes in the boxes have bearing on sustainability (as shown in the circle). All the attributes influence and are influenced by each other.

Table 7.1: Some examples of ecologically sustainable and vulnerable ecosystems

Ecologically sustainable ecosystems	Ecologically vulnerable ecosystems
Natural forests at the ecological climax	Human managed forests
Heterogeneous forests	Monoculture forests
Oak-type forests in temperate climate	Apple orchards
Agroecosystems (forests-cultivated lands based farming systems)	Cultivated croplands
Agroforestry (cropping systems integrated with trees)	Cultivated croplands
Mixed farming systems	Monoculture-based farming systems

Biodiversity, Economic Development, and Sustainability

As intensively discussed in this book, the planet's biodiversity is the basis of survival, material progress, and sustainability of the human race. The biotic component includes all the varieties of plants, animals, and microorganisms that can thrive and all are of vital value for socioeconomic development. The abiotic environment is greatly influenced by the biotic component of the environment, i.e., biodiversity, and vice versa. Substantial alteration in the original composition of the environment leads to its quality deterioration.

Development, speaking in Abraham Lincoln's tune, is the economic well-being of the people, by the people, for the people. Development – in essence – is a process of socioeconomic advancement. This process inevitably takes place through human interaction with environmental resources, primarily nature's biodiversity.

Dependent on nature's diversities, development is a dynamic process. Changes brought about as a result of the development processes cost nature's diversities (ecosystem, species, genetic resources, and physical environment). If the development keeps pace with nature's regeneration capacities, it is a sustainable kind of development. On the other hand, if the development exceeds nature's regeneration capacities, it is a kind of unsustainable development. Sustainable development is characterized by a larger possible resource size, the highest possible level of nature's biodiversity, conservation-oriented changes in resource allocation/ utilization for various purposes and in resource management, environmentally safe technological processes, and social justice. Such a kind of ecologically and socially conscious development is a sustainable development ensuring fulfillment of the needs of the present without compromising the abilities of the future generation to fulfill their own needs. Sustainable development also paves pathways for a happy and sustainable future (Fig. 7.2). If the development processes are resource damaging, biodiversity eroding, haphazard, unwise, and driven by greed rather than by need, then environmental pollution and many other socio-cultural tensions would be the natural outcomes and such a resource-eclipsing development would be an unsustainable one (Fig. 7.3). An unsustainable development would not help us carry on for a long time.

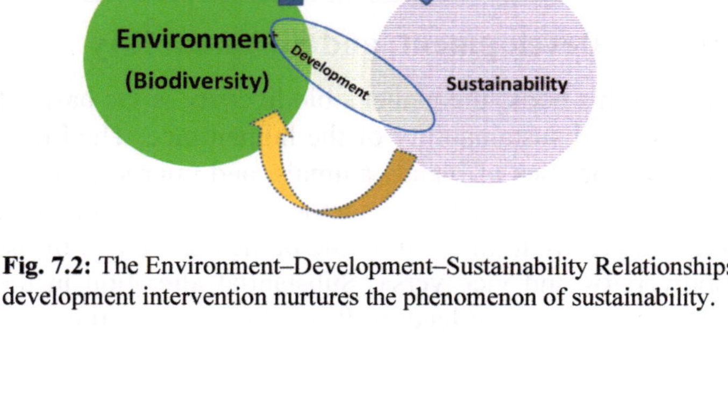

Fig. 7.2: The Environment–Development–Sustainability Relationships: Rational human development intervention nurtures the phenomenon of sustainability.

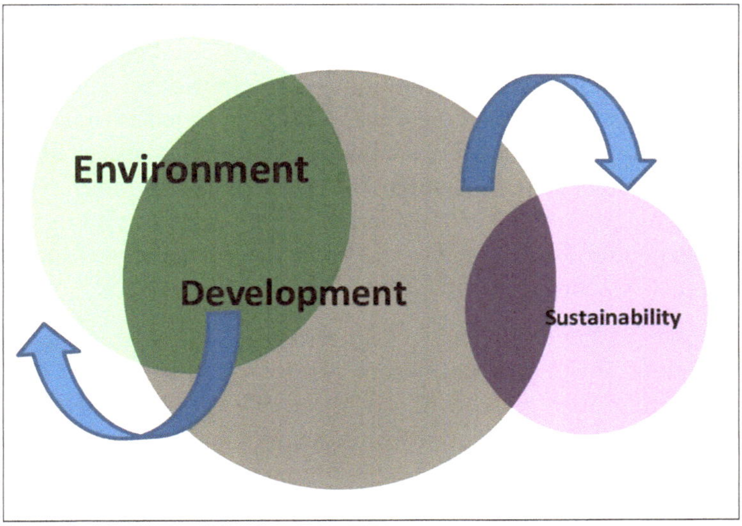

Figure 7.3: When development processes ride over the environment reducing its limits, the state of sustainability breaks down.

Natural Vs Anthropogenic Ecosystems

There are differences in the functioning of a natural ecosystem and that of an artificial or anthropogenic ecosystem. The word "artificial" in the context of ecosystems sounds pretty unnatural. But it is true. Most of the land on Earth bears what we can refer to as man-made, artificial, or anthropogenic ecosystems, for there is substantial human intervention in nature not allowing the ecosystems to evolve the way they could. With continuous, unabated, and ruthless exploitation of the resources of the Earth's ecosystems, and management aimed at extracting maximum benefits from natural ecosystems, most of them are turning into less functional entities of nature than they ought to be when in the natural state of their existence.

When an ecosystem evolves the natural way, at a certain point of its evolutionary time it attains its climax. All natural ecosystems on Earth which are virgin ones are in the state of the climax of their evolution. When in its climax, an ecosystem regulates itself through self-regulating mechanisms. Then it is at the highest level of its functioning, contributing to building up a life-ameliorating microclimate. Anthropogenic ecosystems (such as agricultural systems) – if the human interference is too intensive – gradually advance towards disclimax. At its disclimax, an ecosystem is not as effective in its functioning as it ought to be at its climax.

When in the climax, an ecosystem thrives with maximum photosynthetic efficiency and reverberates with the fullness of life. Intrinsic ecological factors operating within an ecosystem in its climax provide bright chances for new species to evolve over an ecological time scale. Every new species of a living organism has evolved in such climax environments, adding richness to life. In an anthropogenic ecosystem, depending upon the extent of human interference, photosynthetic efficiency would be considerably reduced and paucity and crisis of living forms will prevail. There would be increased chances of species extinction in such ecosystems.

The greater the photosynthetic efficiency of an ecosystem, the higher the rate of primary production. The higher the rate of primary production, the more effective the functioning of an ecosystem. The more effective the functioning of an ecosystem, the higher the productivity of the ecosystem. Thus, the natural or climax ecosystems would attain maximum primary productivity, and the anthropogenic ones, unlike those of the former, will make poor performance. Sometimes even an anthropogenic ecosystem could demonstrate higher productivity provided its native species are altered by monocultures of high-yielding species and external inputs are supplied in plentiful amounts. But, such transformed ecosystems are extremely vulnerable and their productivity is unsustainable.

Biodiversity-Based Sustainability Principles

Biodiversity is not just a bio-physical aspect of life on the living planet. It is also not the 'spice' of life. An outcome of natural evolution itself, it is a phenomenon as well as a crux of life. The processes of life's sustenance go on via biodiversity. The sustainability of life is nurtured basically through nature's biodiversity. There can be no sustainability without biodiversity as its central component. There are three basic principles of sustainability which all revolve around biodiversity:

1. Soil biodiversity
2. Above-soil biodiversity
3. Nutrient Cycles

Soil Biodiversity

As we have discussed in a chapter, the soil is not just a substratum for the plants to anchor their roots. It is a living system—an ecosystem. Biodiversity of soil – or pedobiodiversity (pedo-flora and pedo-fauna) – is the basis of living soil. There is a far greater measure of biodiversity per unit area and volume of the soil than that above the soil visible in the troposphere. The biodiversity of the soil is fed by both chemosynthesis and photosynthesis. It is the living soil from where the structure of the biosphere through the structure of all living beings comes into being. It occurs through nitrogen fixation, especially biological nitrogen fixation. Nitrogen is the fundamental constituent of proteins. Nitrogen in the atmosphere is incorporated into the structure of proteins, via building blocks of the proteins, the amino acids. This phenomenon is attributable to the role of nitrogen-fixing microorganisms in the soil (and also in water). Nitrogen fixed by the microorganisms inhabiting the soils freely and in association with certain plants' root systems is absorbed and assimilated by plants in the form of nitrates.

Soil biodiversity, from the viewpoint of its role in biological nitrogen fixation, would suggest that it is of vital significance even to the above-soil biodiversity. In other words, the above-soil biodiversity itself is nurtured by the soil biodiversity. The soil biodiversity (that, apart from nitrogen-fixing bacteria, includes various types of decomposers) also determines the dynamics of nutrient cycles in the lithosphere (connected with all environmental components). Availability of the essential nutrients in the soil determines the nutrition status – and, therefore, the health – of the above-soil biodiversity. In conclusion, we can say that the level of the above-soil biodiversity (biodiversity in the troposphere) is directly proportional to the level of soil biodiversity. The well-being of the biodiversity visible on the land surface, including all humans, depends on the well-being of the biodiversity below the land surface.

Soil Surface Biodiversity

Above-soil or soil surface biodiversity or the biodiversity visible to us in the troposphere is the biodiversity we generally count for various socioeconomic gains. This biodiversity is the crux of all our management (conservation and utilization) practices count on. The higher the level of the biodiversity above land surface, the greater the degree of sustainability. This biodiversity is fed by photosynthesis. All terrestrial consumers, ranging from herbivores to top carnivores, are dependent on this biodiversity. Entire humanity also directly depends on this biodiversity.

The above-soil biodiversity nourishes the soil biodiversity by supplying organic matter to the soil. Organic matter in the soil is about 2 to 3 times higher than that in the vegetation outside the soil, but most of it comes through photosynthesis by green plants. A small fraction of carbon is contributed through chemosynthesis by microorganisms and photosynthesis by soil algae and blue-green algae. Soil organic matter supports a huge number and high level of diversity of detritivores. The biodiversity of chemosynthetic and photosynthetic organisms, consumers, and detritivores is interconnected; all organisms thrive maintaining interrelationships.

Nutrient Cycles

Nutrients contained in organic matter in the soil are released due to the action of decomposers leading to mineralization. While nitrogen and carbon dioxide get lost in their atmospheric pool, other nutrients – metallic and non-metallic, macro as well as micronutrients – stay in their soil pool. These nutrients become available for the plants above the soil. A proportion of the released nutrients also nourish soil life. The nutrient cycles, both sedimentary and gaseous, in this way, nourish biodiversity within and outside the soil. The nutrient cycles, which on the broader geological and geographical scale or global scale, are referred to as bio-geochemical cycles, are vital for the very ecological integrity of the planet, and the planet's biodiversity plays a key role in maintaining these cycles.

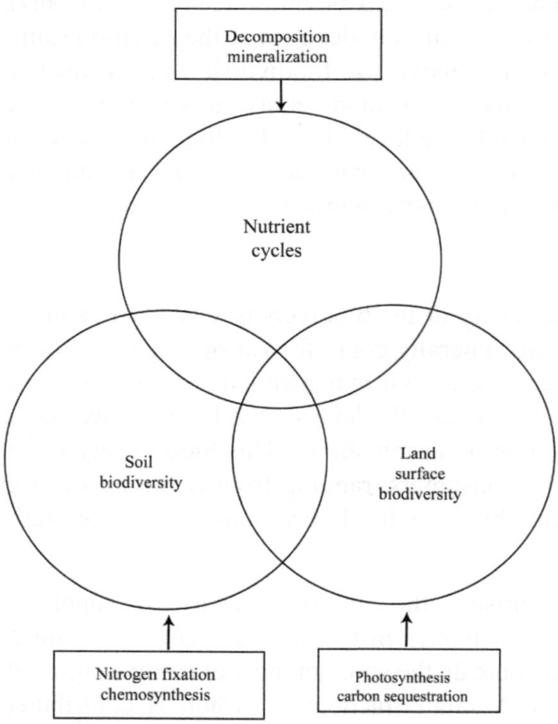

Fig. 7.4: The three interlinked biodiversity-centric sustainability principles: Biodiversity as the main ecosystem structural component performs key ecosystem functions, viz., nitrogen fixation, chemosynthesis, photosynthesis/ carbon sequestration, and decomposition/ mineralization.

The three biodiversity-centric sustainability principles linked with various ecosystem functions are depicted in Fig. 7.4.

Sustainability Management

Natural ecosystems that, in the journey of their evolution, have attained ecological climax are sustainable up to the highest degree of sustainability. Natural ecosystems that, as a result of human intervention, have not been degraded to a large extent are also sustainable to a great extent provided the degradation processes are halted and their ecological regeneration capability is maintained. The need for sustainability management arises in the case of human-managed systems, which are the ecosystems transformed for implementing various socioeconomic activities. If the first and foremost principle of sustainability, that is, ecological sustainability, is not met, we cannot attain a sustainable socioeconomic system, for example, agriculture, animal husbandry, gardening, industry, mining, etc.

In a socioeconomic system, the ecological dimension of sustainability must be as big as it could be able to compensate for ecological damage necessitated for specific purposes, such as a construction activity averting ecological regeneration of a land area. In the case of agriculture, a forest ecosystem serves as a core for establishing a sustainable agroecosystem. Agriculture based on cultivated land and cut off from forests cannot be sustainable. Forest as an integral component of an agroecosystem or a farming system serves as a vibrant and healthy source of nutrients for the cultivated land.

Many insects prospering in a forest play role of pollinators that contribute to pollinating many flowering food crops and, thus, contribute to the ecological integrity of an agroecosystem. A forest is ecologically more stable than cultivated land where food crops with shallow roots depend on the nutrients of the topsoil. In other words, cultivated lands are more fragile than forests. A natural forest with deep-rooted trees as the dominant biotic component naturally embraces a higher level of biodiversity than cultivated lands with a narrow base of biodiversity (agrobiodiversity). Biodiversity is the basis of ecological sustainability. Forest-cultivated land links in agriculture, thus, contribute to enhancing and maintaining the sustainability of the whole system.

Nutrients from a forest ecosystem flow into fragile cultivated land directly through the water during rains, winds, and wild animals, especially birds that feed on forest products (fruits, flowers, insects, etc.). Farmers manage such nutrient flows from forests to cultivated lands by employing livestock. Livestock feed on forest produces, especially tree leaves, and the forest nutrients flow to the cultivated land through their dung/ manure. Livestock also contribute to maintaining cyclic flows of nutrients. They consume crop residues, especially straws and green fodders, and a proportion of the consumed nutrients get recycled into cultivated soils through manure.

Households, which are the major consumers of cultivated produce as well as forest products, also produce a lot of biodegradable/ recyclable wastes. Such wastes contain a bulk of essential plant nutrients. The biodegradable wastes, after being converted into compost/ vermicompost, should be applied to cultivated lands.

Key points of human management of natural resources aimed at realizing sustainability are:

i. Protection and conservation of forests that helps enhance nature's biodiversity;

ii. Increasing agrobiodiversity base in cultivated lands through: a) multiple cropping/ mixed-cropping systems, b) deep-rooted crops combined with shallow-rooted crops, c) increasing genetic diversity of each crop, and d) agroforestry;

iii. Animal husbandry involving multiple livestock species and a variety of breeds of each livestock species;

iv. Waste management: converting organic wastes (dung, food wastes, human wastes) into manure/ compost/ vermicompost and applying the same into cultivated lands that would help enhance soil fertility and soil biodiversity.

An agroecosystem involving natural forest, cultivated land, livestock, and households in integrity with each other and managed for sustainability by operationalizing sustainability principles will be an ecologically sound, environmentally safe, functionally vibrant, and productive agricultural system. The natural forest area in an agroecosystem must be at least 10 times larger than the cultivated land. The natural forest would help maintain soil, water, and biodiversity conservation, moisture circulation, carbon sequestration, and microclimate of an area which are vital for an overall environment conducive to the efficient functioning of the agro-ecosystem. All the functional attributes of a forest ecosystem as an integral component of a sustainable agroecosystem are outlined in Fig. 7.5.

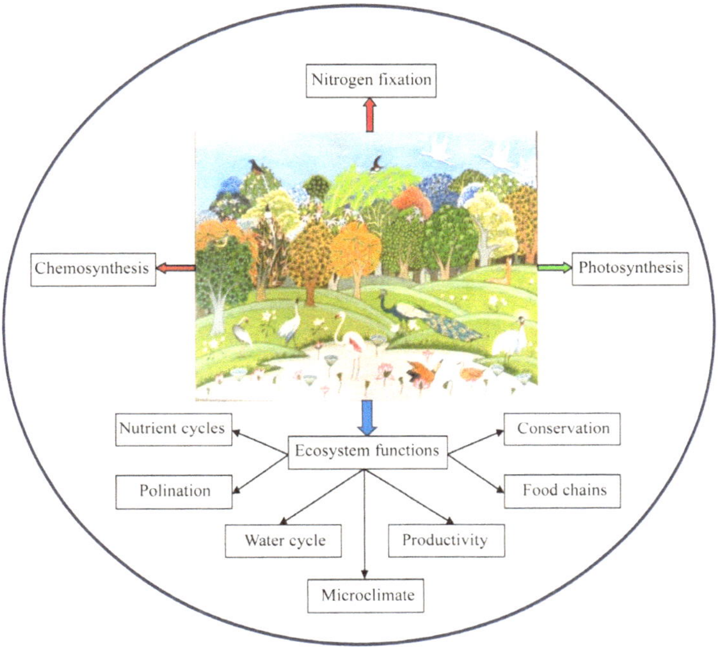

Fig. 7.5: A forest ecosystem as a core component of an agroecosystem has many functional attributes vital for sustainability in agriculture.

Summary

Sustainable Development in the WCED Report has been defined as *Development that meets the needs of the present without compromising the ability of the future generations to meet their own needs*. Biodiversity builds up, induces, enhances, supports, and maintains sustainability and, thus, serves as the basis of environmental sustainability. The higher the level of biodiversity in nature the higher the degree of resilience and the higher the degree of resilience the higher the state of sustainability.

Monocultures introduce vulnerability in ecosystems. These are liable to be damaged by physical factors like strong winds, storms, heavy rains, fires, etc., and by biotic pressures, such as intensive grazing, insects, plant pathogens, etc. Heterogeneous ecosystems, contrarily, are invariably more resilient and, consequently, more sustainable. In other words, the ecosystems with a lesser level of biodiversity are ecologically more vulnerable than the ones with a higher level of biodiversity.

If the development keeps pace with nature's regeneration capacity, it is a sustainable kind of development. On the other hand, if the development exceeds nature's regeneration capacity, it is a kind of unsustainable development. Ecologically sound and socially conscious development is a sustainable development. An unsustainable development would not help us carry on for a long time. There are differences in the functioning of a natural ecosystem and that of an artificial or anthropogenic ecosystem. When in its climax, an ecosystem regulates itself through self-regulating mechanisms. Then it is at the highest level of its functioning, contributing to building up a life-ameliorating microclimate. When in the climax, an ecosystem thrives with maximum photosynthetic efficiency and reverberates with the fullness of life. Intrinsic ecological factors operating within an ecosystem in its ecological climax provide bright chances for new species to evolve over an ecological time scale. The greater the photosynthetic efficiency of an ecosystem, the higher the rates of primary production.

Biodiversity is not just a bio-physical aspect of life on the living planet. It is also not the 'spice' of life. An outcome of natural evolution itself, it is a phenomenon as well as a crux of life. There are three basic biodiversity-centric principles of sustainability which all revolve around biodiversity: i) soil biodiversity, ii) above-soil biodiversity, and iii) nutrient cycles. It is the living soil from where the structure of the biosphere through the structure of all living beings comes into being. It occurs through nitrogen fixation, largely through biological nitrogen fixation. Availability of the essential nutrients in the soil determines the nutrition status – and, therefore, the health – of the above-soil biodiversity. The level of the above-soil biodiversity (biodiversity in the troposphere) is directly proportional to the level of soil biodiversity. The above-soil biodiversity is fed by photosynthesis. All terrestrial consumers, ranging from herbivores to top carnivores, are dependent on this biodiversity. Entire humanity also directly depends on this biodiversity. Nutrient cycles, both sedimentary and gaseous, nourish biodiversity within and outside the soil and are vital for the very ecological integrity of the planet. The planet's biodiversity plays a key role in maintaining these cycles.

Natural ecosystems have attained ecological climax are sustainable up to the highest degree of sustainability. Natural ecosystems that have not been degraded to a large extent are also sustainable to a great extent provided the degradation processes are halted and their ecological regeneration capability is maintained. The need for sustainability management arises in the case of human-managed systems, which are the ecosystems transformed for implementing various socioeconomic activities. If the first and foremost principle of sustainability, that is, ecological sustainability, is not met, we cannot attain a sustainable socioeconomic system such as agriculture, animal husbandry, gardening, industry, mining, etc.

Forest as an integral component of an agroecosystem or a farming system serves as a vibrant and healthy source of nutrients for the cultivated land. Many insects prospering in a forest play role of pollinators that contribute to pollinating many flowering food crops and, thus, contribute to the ecological integrity of an agroecosystem. A natural forest with deep-rooted trees as the dominant biotic component naturally embraces a higher level of biodiversity than cultivated lands with a narrow base of biodiversity (agrobiodiversity). Livestock also contribute to maintaining cyclic flows of nutrients. Households also produce a lot of biodegradable/ recyclable wastes. The biodegradable wastes, after being converted into compost/ vermicompost, should be applied to cultivated lands. Key points of human management of natural resources aimed at realizing sustainability are: i) Protection and conservation of forests; ii) increasing agrobiodiversity base in cultivated lands; iii) harnessing the role of livestock; and iv) waste management. An agroecosystem involving natural forest, cultivated land, livestock, and households in integrity with each other and managed for sustainability by operationalizing sustainability principles will be an ecologically sound, environmentally safe, functionally vibrant, and sustainably productive agricultural system.

Exercises

A. Multiple Choice Questions

1. The term 'sustainable development' was first of all used in 1980 in the
    a. World Conservation Strategy
    b. World Commission on Environmental and Development
    c. UN Commission on Sustainable Development
    d. FAO's Strategies for Climate Change Mitigation
2. The term 'Sustainable Development' was first of all defined and elaborated in the report of
    a. WCED b. IPCC
    c. WMO d. FAO
3. What is the title of the Report in which 'Sustainable Development' has been defined and elaborated?
    a. Our Common Future b. Sustainable Development
    c. Environment and Development d. Man and Biosphere

4. The Report in which 'Sustainable Development' has been defined and emphasized was popularly known as the

   a. Sustainable Development Report b. Brundtland Report

   c. Kurt Waldheim Report d. Nixon Report

5. Which of the following statements is wrong?

   a. There can be no sustainability without ecological sustainability of its own.

   b. A sustainable economy cannot be ascertained without ecosystem considerations.

   c. Economic growth is the first and foremost dimension of sustainability.

   d. All sorts of livelihood and economic systems ultimately survive on and function and sustain through ecosystem functioning.

6. Which of the following statements is correct?

   a. A vibrant and sustainable economy can be ensured only through industrialization.

   b. A healthy and vibrant ecosystem is not a precondition for realizing sustainability.

   c. Nature's biodiversity is the basis of environmental sustainability.

   d. The processes of sustainability are independent of biodiversity.

7. The greater the photosynthetic efficiency of an ecosystem,

   a. The lower the rates of oxygen production.

   b. The lower the rates of carbon sequestration.

   c. The higher the rates of primary production and carbon sequestration.

   d. The higher the rates of CO_2 release into the atmosphere.

8. An agroecosystem comprises a

   a. Forest b. Cropland

   c. Livestock d. All of the above

9. Which is the core component of an agroecosystem?

   a. A forest b. Cropland

   c. Livestock d. Households

10. What of the following would represent the highest degree of sustainability?
    a. A pond
    b. A forest at its ecological climax
    c. A partially degraded forest
    d. Cropland with the wheat crop
11. An example of pedo-fauna is
    a. Pigeon b. Earthworm
    c. Blue-green algae d. Mushroom
12. Most nitrogen fixation in nature takes place through
    a. Thundering during rainfall (physical) b. Microorganisms
    c. Industrial means d. None of the above
13. Inorganics in the environment are converted into organics through
    a. Photosynthesis b. Chemosynthesis
    c. Photosynthesis and chemosynthesis d. None of the above
14. Most of the organic matter in the biosphere occurs in
    a. Plants b. Animals
    c. Microorganisms d. Soils
15. Which of the following ecosystems is likely to comprise the highest level of biodiversity per unit mass and per unit volume?
    a. Forest b. Grasslands
    c. A lotic ecosystem d. Fertile soil
16. What is true about a forest ecosystem?
    a. Many insects prospering in a forest play role of pollinators.
    b. A natural forest with deep-rooted trees as the dominant biotic component naturally embraces a higher level of biodiversity than cultivated lands with a narrow base of biodiversity (agrobiodiversity).
    c. Livestock play a crucial role in nutrient recycling in an agroecosystem.
    d. All of the above.

17. What would be more sustainable than the other?
    a. An oak-type forest more than an apple orchard
    b. Monocultures forests more than biodiversity-laden natural forests
    c. An agricultural system based on croplands than an agroforestry system
    d. Shallow-rooted crop annuals than deep-rooted perennials
18. Monocultures
    a. Stabilize ecosystems
    b. Introduce vulnerability in ecosystems
    c. Enhance nutrient cycles
    d. Do not affect an ecosystem's sustainability
19. Biodiversity
    a. Has no connection with ecosystem resilience
    b. Decreases an ecosystem's resilience
    c. Enhances an ecosystem's resilience
    d. May decrease or enhance an ecosystem's resilience
20. The relationship of resilience with sustainability is
    a. Negative
    b. Neutral
    c. Negative or positive
    d. Positive

Answers

1. **a**, 2. **a**, 3. **a**, 4. **b**, 5. **c**, 6. **c**, 7. **c**, 8. **d**, 9. **a**, 10. **b**, 11. **b**, 12. **b**, 13. **c**, 14. **d**, 15. **d**, 16. **d**, 17. **a**, 18. **b**, 19. **c**, 20. **d**

B. Short-Answer Type Questions

1. What do you mean by sustainability?
2. What is the difference between sustainable development and sustainability?
3. What is central to the process leading to sustainability?
4. Give some examples of ecologically sustainable and vulnerable ecosystems.
5. What is the difference between natural and anthropogenic ecosystems?

6. Give some examples of natural and anthropogenic ecosystems.
7. What are the various dimensions of sustainability?
8. How is socioeconomic sustainability rooted in ecological sustainability?
9. What do you know about monocultures?
10. Why are the systems based on monocultures ecologically vulnerable?

C. Long-Answer Type Questions

1. How biodiversity is linked with sustainability?
2. What are the various functional attributes of a forest in an agroecosystem? How is the agroecosystem-based agriculture more sustainable than the one based on cultivated land?
3. What are the principles of sustainability? Describe each revealing their interrelationships.
4. How can biodiversity-centric sustainability be operationalized?
5. Write short notes on the following:
   a. Biodiversity-Sustainability Links
   b. Soil Biodiversity
   c. Role of Livestock in an Agroecosystem

References

IUCN-UNEP-WWF. 1980. World Conservation Strategy: Living Resource Conservation for Sustainable Development. Gland: Switzerland, IUCN. 77pp.

Singh, Vir. 2019. Fertilizing the Universe: A New Chapter of Unfolding Evolution. London: Cambridge Scholars Publishing. 285pp.

Singh, Vir. 2020. Environmental Plant Physiology: Botanical Strategies for a Climate Smart Planet. Boca Raton" Taylor and Francis (CRC Press). 230 pp.

Web Resources

https://onlinelibrary.wiley.com/doi/abs/10.1002/9781119434016.ch12

https://sustainability.stackexchange.com/questions/476/what-is-the-relationship-between-biodiversity-and-sustainability

https://sustainability-academy.org/ecological-footprint/

https://sustainabledevelopment.un.org/topics/biodiversityandecosystems

https://unfoundation.org/blog/post/biodiversity-essential-sustainable-development/

https://www.footprintnetwork.org/

https://www.ijser.org/researchpaper/Biodiversity-and-Sustainable-Development-A-Critical-Analysis.pdf

https://www.indiascienceandtechnology.gov.in/world-environment-day/biodiversity-and-sustainable-development-goals

https://www.nature.com/articles/s41893-019-0417-9

Glossary

Adaptation an evolutionary process of creating anatomical, physiological, and/ or behavioral changes enhancing the abilities of a population to thrive in a particular environment.

Alfa diversity the diversity within a community, that is, the diversity of the organisms sharing a common community.

Autotroph an organism with the ability to synthesize its food (organic molecules) using CO_2 and H_2O and energy from sunlight (photoautotroph) or inorganic molecules like H_2S (chemoautotroph).

Beta Diversity the diversity between communities.

Biosphere portion of the Earth that supports life, or the total of all ecosystems of the Earth, or the total global ecosystem.

Carbon sequestration the mechanism that involves the removal of atmospheric CO_2 to be captured, secured, stored elsewhere, and/ or used for some specific purpose. In the context of climate change mitigation, the purpose of carbon sequestration is to strike a global carbon balance so that the enhanced global warming effect of CO_2 could be nullified or minimized.

Carnivore an animal that feeds on other animals (approximately synonymous with a predator).

Chemosynthesis the synthesis of organic compounds by bacteria (or other organisms) involving energy released from the reaction of inorganic compounds, typically in the absence of sunlight.

Chemosynthetic pertains to autotrophs that use inorganic molecules as a source of energy in the synthesis of organic compounds.

Climax community a community emerging in late succession which is stable until disrupted by disturbance.

Community all of the interacting organisms living in a particular area or space called an ecosystem or the populations of all interacting species in an ecosystem.

Community structure attributes of a community, e.g. number of species, the distribution pattern of individuals among the species within the community.

Composting the process that involves the degradation of organic wastes, especially animal excreta, by microorganisms in the presence of oxygen.

Decomposition a key process in nutrient cycling involving the breakdown of organic matter with the release of constituent components, e.g. CO_2, H_2O, minerals, etc.

Density the number of individuals in a population per unit area or volume, e.g. number of plants per m^2, number of fish per m^3 of water, and number of people per km^2.

Desert an arid biome spread over about 20% of land surface with evapotranspiration rates resulting in water loss far higher than precipitation for most of the year.

Detritivore refers to organisms that feed on non-living organic matter.

Distribution the spatial arrangement of individuals in a local population, or the natural geographic range of an organism.

Drought an extended period of dry weather, or insufficient periodic precipitation, leading to crop damage, ecosystem functioning impairment, and paucity of drinking water for human and animal populations.

Ecological sustainability the very basis of all levels, all kinds, and all dimensions of sustainability. Sustainability, in essence, is rooted in ecological sustainability.

Ecology study of reciprocal relationships between organisms and the environment.

Ecosystem a biological community along with all the abiotic factors influencing that community.

Endemic a term used for the species and populations thriving only in a locality and not anywhere else.

Equilibrium a state of (perfect) balance with opposing factors canceling each other.

Evolution a natural phenomenon that brings changes in the populations of organisms over time; it generally involves changes in the gene frequencies in a population.

Exploitation an interaction among species in which the fitness of the individuals of the exploiting species (e.g. the predators, the parasites, etc.) gets enhanced over the exploited ones (e.g. the host, the prey, etc.). It also pertains to the unwise and excessive use of natural resources, ecosystems, and organisms by human beings.

Fecundity the number of seeds produced by a plant or of eggs produced by an animal.

Food chain a sequence of matter and energy transferred as food from organism to organism.

Food web a network of food chains interconnected at various trophic levels and forming several feeding connections among all the members of a living community.

Gamma diversity the diversity of the communities or habitats over a geographical area or the entire landscape.

Herbivore a plant-eating heterotrophic organism.

Heterotroph an organism using organic molecules as a source of nutrients and energy; it may be a herbivore, a carnivore, an omnivore, or a detritivore.

Hydrological cycle the sun-driven cycle of water through the biosphere involving evaporation, transpiration, condensation, precipitation, and runoff.

IUCN Red List a list of threatened species; also called Red Data List, it is the world's most comprehensive inventory of biological species.

Keystone species the species that, despite low biomass, play a crucial role in affecting many other species and, thus, community structure.

Landform any distinctive feature of the Earth's surface.

Landscape an area of land with a patchwork of ecosystems.

Macroclimate the general climate prevailing in a region.

Mangrove a forest of tropical and subtropical marine shores that is dominated by salt-tolerant woody species such as *Avicennia* and *Rhizophora.*

Microclimate variation in general climate (macroclimate) at a small scale due to specific local features, such as aspect, substrate, vegetation, etc.

Mineralization the breaking down of organic matter into inorganic constituents during the process of decomposition.

Mycorrhizae a mutualistic association between fungi and plant roots.

Natural selection differential reproduction and survival of the individuals in a population under the influence of the environment; proposed by Charles Darwin as a mechanism driving evolution.

Niche the environmental factors influencing the survival, growth, and reproduction of a species.

Noise an undesirable and unwanted sound that becomes an irritant and a source of stress and often of some diseases.

Nutrient a substance (a constituent of the food) that nourishes an organism vital for its survival, growth, maintenance, reproduction, and evolution.

Nutrient cycle the use, transformation, movement, and reuse of a nutrient in the ecosystems.

Ocean anoxia a condition of oxygen depletion in the oceans; the ongoing global warming may induce this condition by gradually reducing the water-dissolving properties of the oceans.

Oligosaprobic a term used for the organisms inhabiting clean water with high dissolved oxygen, such as algal species, namely *Ulothrix, Cladophora, Ulothrix, Bulbochaete*, and diatoms and animals species, like *Vorticella, Planaria, Notholca*, etc.

Oligotrophic a term used for characterizing a lake with low nutrient content, high oxygen level, and lower primary productivity.

Parasite an organism that thrives on or in another organism (s) (the host) and derives benefits for itself; it reduces the fitness of the host but does not kill it.

Pathogen an organism – commonly among viruses, bacteria, and protozoa – that induces disease, a debilitating condition, in its host.

Peroxyacyl nitrates (PANs) a family of phytotoxins, with PAN as the most abundant member of the family, in the photochemical smog, causing damage to chloroplasts and consequently impairing photosynthesis, electron transport chain of respiration, enzyme activity, and anabolic processes in plants.

Photosynthesis the process in which chlorophyll molecules in plants, algae, and bacteria absorb light and transfer their energy to electrons; the energy by these electrons is used to synthesize ATP and NADPH, which in turn serve as donors of electrons and energy for the synthesis of sugars.

Photosynthetic about the organisms (green plants, algae, cyanobacteria) capable of photosynthesis.

Pollution an undesirable (or negative) change in the physical, chemical and biological characteristics of the environment or of a component of the environment (air, water, soil, etc.) occurring directly and/ or indirectly that may affect life adversely in one way or the other or may carry potential to harmfully affect life instantly or over some time.

Population a group of individuals of a species in an ecosystem/ community.

Predator a heterotrophic organism, usually a carnivore, that kills other organisms for food.

Prokaryotes the organisms with cells having no membrane-bound nucleus or organelles; e.g., bacteria and archaea.

Radiation the transfer of heat through electromagnetic radiation.

Resilience the capacity, generally of an ecosystem or a community, to recover its structure and functioning after disturbance.

Resistance the capacity, generally about an ecosystem or a community, to maintain its integrity, including its structure and functioning, despite facing potential disturbance.

Sacred forests or sacred groves represent an alternative approach to conservation that involves the principles emanating from a diversity of prevailing cultures or social taboos rather than the approaches based on the legal framework.

Species diversity a measure of diversity at the species level; it increases with species evenness and species richness.

Species evenness denotes the relative abundance of a species in a community.

Species richness the number of species in a community.

Stability the persistence of an ecosystem or a community in the face of disturbance; an attribute of the combined effect of resistance and resilience.

Stress an environmental condition, such as the one generated by excessive heat, cold, radiation, and a pollutant, that negatively influences the physiology of the organisms and their performance (e.g., growth rate, reproduction, lactation, etc.) and/ or alters structure and functioning of an ecosystem.

Succession the gradual change in the communities in an area beginning with a bare rock (primary succession) or following a disturbance (secondary succession).

Sustainability although often used as synonymous with sustainable development, sustainability generally refers to the human ability to live constantly in tune with the biosphere's capacities.

Sustainable development the development that meets the needs of the present without compromising the ability of the future generations to meet their own needs (*Our Common Future*, 1987).

Sustainable Future a future that would be witnessed by healthy, happy, and vibrant humanity (Singh 2019).

Temperate forests the deciduous or coniferous forests prevailing amid the temperate climatic regions of the world, mostly between 40° to 50° latitudes.

The Sixth Mass Extinction an ongoing species extinction event during the present Anthropocene or Holocene epoch.

Trophic dynamics the transfer of energy from one part or trophic level of an ecosystem to another.

Trophic level the trophic (feeding) position in an ecosystem/ community constituted by, e.g., producers, herbivores, and carnivores.

Tropical areas the areas near the equator and between the Tropics of Cancer in the northern hemisphere and Tropic of Capricorn in the southern hemisphere of the globe with a warm to hot and moist climate year-round.

Troposphere the lowermost atmospheric layer extending from the Earth's surface to an altitude of 9 to 16 km and is characterized by a decrease in temperature with increasing altitude.

Tundra a northern biome dominated by mosses, lichens, and dwarf willows prevailing in extremely cold and harsh environments and having a very short growing season.

Vermicomposting or Vermiculture the process that converts solid organic wastes, such as kitchen wastes, sludge, and dung, using earthworms into nutrient-rich vermicompost used as high-quality organic fertilizer in organic farming.

Water cycle (see hydrological cycle).

Zooplankton the animals, mostly microscopic, that drift in the surface waters of the lakes or oceans.

Subject Index

10% Law 80, 90

A

agriculture 37
  and sustainable future 37
  gene revolution 30, 31, 52, 55, 56
  Green Revolution 27, 28, 29, 30, 31, 33, 34, 36, 37, 38, 39, 52, 55, 56, 70, 142,
  LPG 27, 29, 30, 37, 38, 39, 40, 52, 53, 55, 56
  mainstream 40, 41, 42, 57
  mountain 9, 10, 11, 40, 41, 42, 43, 44, 45, 46, 47, 48, 49, 50, 53, 54, 56, 57, 68, 120
  primitive 27, 28, 3035, 36, 37, 38, 39, 51, 52, 53, 55, 56, 57, 137,
  pyramid 34, 35, 36, 37, 38, 39, 53, 55, 58, 85, 86, 93
  sustainable 6, 25, 26, 28, 30, 35, 36, 37, 38, 43, 46, 53, 115, 116, 118, 120, 125, 126, 137, 138, 139, 140, 147, 148, 149, 150, 155, 157, 158, 159, 160, 162, 163, 170
  traditional 15, 27, 28, 29, 30, 33, 36, 37
agrobiodiversity 25, 26, 27, 28, 29, 30, 31, 32, 35, 36, 38, 39, 42, 52, 53, 55, 57, 100, 101, 155, 156, 159, 161
agroecosystem 40, 42, 43, 44, 45, 46, 49, 50, 53, 54, 56, 57, 58, 155, 156, 157, 159, 160, 161, 163
  diversity 1, 2, 3, 4, 5, 6, 7, 8, 10, 14, 15, 18, 19, 20, 21, 23, 26, 27, 28, 29, 33, 34, 37, 40, 41, 42, 43, 46, 52, 54, 56, 60, 61, 63, 71, 89, 115, 116, 117, 120, 128, 129, 131, 138, 139, 148, 154, 156
  integrity of 44, 47, 118, 154, 155, 158, 159,
  mountain 9, 11, 40, 41, 42, 43, 44, 45, 46, 47, 48, 49, 50, 53, 54, 56, 57, 120,
  sustainable 6, 25, 26, 28, 30, 35, 36, 37, 38, 43, 46, 53, 115, 116, 118, 120, 125, 125, 126, 137, 138, 139, 140, 147, 148, 149, 150, 155, 156, 157, 158, 159, 160, 162, 163
alpha diversity 5, 18, 21, 23
alpine 10, 11, 22, 18
  meadows 48
  vegetation in 11
animals 1, 2, 4, 8, 9, 14, 15, 16, 18, 21, 25, 26, 29, 31, 46, 47, 48, 52, 55, 60, 62, 67, 68, 72, 73, 74, 76, 82, 85, 89, 93, 97, 98, 100, 101, 102, 107, 108, 117, 122, 124, 127, 128, 133, 139, 149, 155, 161
anthropocene 87, 170
archaea 25, 55, 81, 82, 88, 169
archaebacteria 85, 89
autotrophs 25, 57, 64, 65, 79, 80, 82

B

bacteria 3, 25, 31, 37, 43, 54, 55, 62, 63, 64, 65, 72, 73, 76, 81, 82, 83, 84, 85, 88, 89, 91, 93, 153
  arcobacter 83, 88
  beggiatoa 83, 88,
  gammaproteobacteria 82, 83, 88
  sulfur-oxidizing 83. 84, 85, 88, 89, 91
  iron-oxidizing 82, 88
  methanogenic 82, 88
  thiothrix 83, 88, 83, 88,
  thiomargarita 83, 88
beta diversity 5, 18, 21, 23
biodiversity 1, 2, 3, 4, 6, 7, 8, 9, 10, 12, 16, 17, 18, 19, 22, 23, 25, 26, 27, 28, 29, 30, 32, 33, 34, 35, 39, 40, 41, 43, 45, 46, 52, 53, 57, 59, 60, 61, 62, 65, 69, 70, 71, 72, 74, 76, 77, 79, 80, 81, 83, 86, 87, 88, 89, 90, 91, 92, 93, 94, 97, 98, 99, 100, 101, 103, 104, 106,

107, 108, 109, 110, 111, 112, 113, 115, 116, 117, 118, 125, 126, 128, 129, 130, 131, 132, 134, 135, 136, 137, 138, 139, 140, 141, 143, 144, 147, 148, 149, 150, 152, 153, 154, 155, 156, 157, 158, 159, 160, 161, 163
COB 87, 92
community 1, 3, 4, 5, 6, 7, 8, 9, 16, 18, 19, 20, 21, 23, 43, 44, 45, 54, 59, 62, 71, 72, 74, 79, 81, 83, 84, 85, 86, 89, 83, 93, 98, 103, 108, 115, 128, 129, 130
concepts 1
conservation 1, 3, 9, 17, 26, 31, 32, 43, 52, 59, 71, 74, 77, 87, 89, 92, 99, 103, 106, 109, 112, 115, 116, 117, 118, 120, 122, 125, 126, 128, 129, 131, 132, 133, 134, 136, 137, 138, 139, 140, 141, 143, 144, 147, 150, 153, 156, 159
Convention on 59, 87, 92, 117, 118, 139, 141
crises 69, 74, 86, 89
ecosystem 3, 4, 5, 6, 7, 8, 9, 17, 18, 19, 21, 23, 40, 43, 44, 46, 47, 48, 60, 62, 63, 66, 67, 69, 71, 72, 74, 82, 85, 91, 98, 99, 103, 105, 108, 116, 117, 118, 125, 126, 138, 148, 150, 151, 152, 154, 155, 156, 157, 158, 160, 161, 162
genetic 1, 2, 3, 4, 9, 15, 16, 18, 20, 21, 23, 26, 28, 29, 32, 33, 37, 52, 54, 97, 101, 107, 115, 120, 126, 131, 138, 140, 141, 150, 156
gradients of 8, 23
hotspots of 134, 136, 140, 144
latitudinal effects on 8
levels of 3, 9, 23, 59, 72, 85, 148
species 1-14, 16-21, 23, 26-29, 31- 34, 37, 38, 48, 51-55, 57, 60-62, 65-68, 72, 73, 75, 79, 83-89, 92, 93, 97-101, 103-113, 115-118, 120, 122, 123, 128-134, 139, 140, 148, 150, 152, 156, 158
threats to 86, 94, 97, 107, 113
values of 13, 14, 19, 23, 32
biogeographical classification 10
biosphere reserve 122, 125, 126, 127, 128, 139, 142, 143, 144
botanical garden 132, 133, 143, 144
Brundtland Commission 137, 147
Brundtland Report 160
Our Common Future 137, 147, 159, 170

C
carnivory 25, 54, 106
chemolithoautotrophy 25, 54, 82
chemosynthesis 25, 62, 65, 72, 79, 81, 82, 83, 84, 85, 86, 87, 88, 89, 91, 92, 93, 153, 154, 161
communities 1, 5, 6, 9, 18, 21, 41, 59, 60, 72, 79, 80, 81, 83, 85, 86, 88, 89, 90, 93, 104, 109, 110, 116, 120, 125, 129, 130, 131, 139
chemosynthesis-based 62, 72, 79, 81, 84, 86, 88, 89, 93
photosynthesis-based 79, 88, 93
community diversity 5, 6, 18, 21, 23
conservation 1, 3, 9, 17, 26, 31, 32, 43, 52, 59, 71, 74, 77, 87, 89, 92, 99, 103, 106, 109, 115, 116, 117, 118, 120, 122, 125, 126, 128, 129, 131, 132, 133, 134, 137, 138, 139, 140, 141, 143, 144, 147, 150, 153, 156, 159
ex-situ 112, 116, 120, 131, 138, 139, 140, 141, 144
in-situ 116, 120, 125, 129, 138, 139, 140, 144
strategies for 116, 117, 159
crops 3, 14, 15, 19, 28, 29, 31, 32, 33, 35, 36, 37, 39, 42, 46, 47, 48, 49, 53, 56, 71, 100, 101, 102, 155, 156, 159
Bt 29, 56
commercial 14, 19, 32, 86, 98, 107
deep-rooted 46, 70, 155, 156, 159, 161, 162
food 3, 6, 9, 15, 17, 25, 26, 28, 29, 30, 31, 32, 33, 34, 35, 37, 38, 43, 44, 45, 46, 47, 50, 52, 53, 56, 57, 68, 67, 71, 72, 80, 81, 90, 98, 99, 100, 101, 106, 138, 155, 156, 159
food grain 15, 28, 29, 99
shallow rooted 46, 71, 155, 156, 162
cultivated land 27, 28, 38, 43, 45, 47, 48, 50, 52, 53, 97, 100, 155, 156, 159, 163,
cryopreservation 131, 132, 140, 142, 144
cyanobacteria 62, 63, 65, 72, 73, 75, 76, 80, 90

D
detritus 66, 73, 76, 80, 81, 90
drugs and medicine 15

E

Earth Summit 59, 117, 137, 140
earthworms 43, 55, 62, 68, 72, 73, 77
economic development 125, 149
ecosystem 3, 4, 5, 6, 7, 8, 9, 17, 18, 19, 20, 21, 23, 40, 43, 44, 46, 47, 48, 60, 62, 63, 66, 67, 69, 71, 72, 74, 82, 85, 91, 98, 99, 103, 105, 108, 116, 117, 118, 125, 126, 138, 148, 150, 151, 152, 154, 155, 156, 157, 158, 160, 161, 162
  aquatic 2, 18, 31, 59, 68, 79, 80, 81, 88, 90, 99, 100, 101, 130, 131
  soil 9, 10, 17, 18, 22, 29, 33, 73, 44, 45, 46, 49, 50, 59, 60, 61, 62, 63, 64, 65, 66, 67, 68, 69, 70, 71, 72, 73, 74, 75, 76, 77, 79, 81, 88, 99, 101, 103, 152, 153, 154, 156, 158, 161, 163
  structure 1, 4, 6, 46, 60, 72, 82, 85, 93, 153, 158
  terrestrial 2, 18, 31, 58, 59, 60, 61, 62, 68, 71, 78, 79, 81, 88, 90, 97, 100, 101, 119, 122, 134, 136, 140, 153, 158
ecosystem diversity 3, 5, 6, 7, 8, 18, 19, 20, 21
ecosystem services 17
energy 9, 14, 17, 26, 28, 34, 35, 36, 37, 39, 42, 44, 45, 48, 49, 50, 53, 64, 65, 66, 72, 73, 75, 79, 80, 82, 83, 84, 85, 88, 89, 90, 91, 93, 100, 104, 105, 106
environment 3, 7, 9, 16, 19, 29, 33, 37, 41, 44, 50, 54, 59, 60, 65, 67, 72, 73, 84, 89, 98, 101, 108, 115, 125, 129, 133, 137, 147, 149, 150, 151, 156, 159, 161
environmental pollution 29, 101, 113, 150
eubacteria 85, 89
exotic species 98, 99, 108, 110, 113, 142
extinction of species 103, 113
  anthropogenic extinction 104, 105, 108, 109, 111
  causes of extinction 103
  mass extinction 87, 104, 105, 108, 109, 111
  natural extinction 103, 104, 108, 111
  susceptibility to extinction 105, 106, 111, 113

F

FAO 32, 33, 53, 61, 76, 138, 140, 159
food crops 15, 32, 36, 38, 47, 101, 155, 159
food production 28, 30, 31, 33, 34, 35, 38, 43, 44, 46, 50, 53, 57, 100
forest 9, 11, 12, 28, 38, 42, 43, 45, 46, 47, 48, 52, 55, 61, 69, 70, 71, 72, 74, 97, 102, 120, 121, 122, 129, 136, 141, 147, 155, 156, 157, 159, 160, 161, 162, 163
fungi 1, 25, 31, 43, 55, 60, 62, 63, 64, 65, 72, 73, 75, 76

G

Galápagos Rift 83, 85, 92
gamma diversity 5, 6, 18, 21, 23, 41
genetic diversity 3, 4, 18, 20, 21, 23, 26, 28, 29, 33, 37, 52, 120, 131, 156
genetically modifies organisms 29
giant tubeworm 54, 84, 89, 91, 92, 93, 94
  Riftia pachyptila 84, 8589, 92, 94
GMOs 29
gradients of biodiversity 8, 23
grass 15
grassland 28, 42, 45, 52, 67

H

habitats 2, 6, 8, 10, 15, 18, 32, 61, 67, 68, 97, 98, 101, 102, 107, 108, 113, 116, 118, 120, 122, 130, 131, 133, 138, 139, 140, 167
herbivory 25
heterotrophs 14, 19, 25, 55, 57, 59, 62, 64, 65, 72, 73, 79
Himalayas 10, 38, 41, 45, 47, 49, 50, 51, 52, 54, 57, 48, 99, 136, 137
holocene (see also anthropocene)
households 45, 49, 50, 156159, 160
hydrothermal vents 18, 22, 79, 81, 83, 84, 85, 87, 87, 88, 89, 92, 93

I

Indo-Burma 136, 137
international obligations 87
ISA 87, 92
IUCN 105, 160, 107, 109, 112, 113, 117, 118, 120, 122, 138, 139, 140, 147
IUCN Red List 106, 107, 109, 112, 167
  critically endangered 107, 109, 112
  data deficient 107, 109, 112
  endangered 11, 100, 106, 107, 108, 109,

110, 112, 123, 129, 130, 133, 136
extinct 33, 53, 97, 103, 105, 107, 108, 109, 111, 112, 113
extinct in the wild 103107, 112, 113
least concern 107
near threatened 107
not evaluated 107, 109, 112
threatened 99, 100, 106, 107, 108, 109, 112, 133, 134, 136, 140
vulnerable 3, 36, 43, 86, 101, 105, 107, 109, 110, 112, 117, 130, 148, 149, 152, 157, 162, 163

K
keystone species 7, 19, 21, 85, 113, 167

L
livestock 28, 29, 35, 36, 38, 42, 43, 44, 45, 46, 47, 48,49, 53, 54, 58, 101, 108, 138, 140, 155, 156, 159, 160, 161, 163
living soil 60, 62, 72, 74, 152, 153, 158

M
mainstream agriculture 71, 42, 57
Man and Biosphere 125, 139, 143, 159
meadows 48
microbes (see also microorganisms) 82, 84
microorganisms (see also microbes) 1, 2, 4, 9, 14, 17, 18, 26, 31, 52, 60, 63, 72, 81, 82, 88, 98, 99, 107, 149, 153, 161
monera 1, 60
mountain agriculture 41, 42, 46, 48
mountain agroecosystem 40, 43, 44, 56, 57
accessibility 40, 41, 53, 56, 87
adaptation mechanisms 40, 41, 53, 56
ecological niche 20
fragility 40, 41, 42, 53, 56
heterogeneity 5, 40, 41, 42, 53, 56
marginality 40, 41, 42, 53, 56, 57

N
national park 118, 120, 121, 122, 128, 139, 141, 142, 143
natural resources 7, 17, 19, 29, 37, 100, 105, 116, 118, 138, 156, 159, 167
nutrient cycles 6, 17, 70, 152, 153, 154, 158, 162
NOAA 85, 123

P
pastures 12, 22
pedobiodiversity 60, 152
pedodiversity 60, 61, 77
pedosphere 59, 60, 62, 64, 71, 72, 74, 77
photosynthesis 16, 25, 26, 35, 39, 62, 63, 72, 75, 80, 81, 83, 84, 88, 90, 91, 93, 94, 106, 153, 154, 158, 161, 168
plain agriculture
plants 1, 2, 4, 8, 9, 11, 14, 15, 16, 17, 18, 19, 21, 26, 28, 31, 32, 33, 34, 35, 38, 39, 44, 46, 50, 52, 53, 54, 55, 57, 58, 60, 61, 65, 66, 73, 76, 80, 82, 90, 98, 99, 100, 101, 115, 129, 131, 132, 133, 134, 137, 139, 143, 148, 149, 152, 153, 154, 161
pollution 29, 44, 101, 108, 113, 131, 150, 169
prokaryotes 25, 55, 169
protists 25, 55, 85, 89
pyramid 34, 35, 36, 37, 38, 39, 53, 55, 58, 85, 86, 93

R
rangeland 28, 52
resilience 3, 17, 36, 36, 42, 43, 46, 148, 157, 162

S
sacred groove 127, 130, 131, 169
sacred lake 130
sacred trees 31, 128, 130
Sixth Mass Extinction 87, 109, 170
soil 9, 10, 17, 22, 29, 33, 43, 44, 45, 46, 46, 49, 50, 59, 60, 61, 62, 63, 64, 65-77, 79, 81, 88, 99, 101, 103, 130, 152, 153, 154, 156, 158, 161, 163
soil biodiversity 29, 59, 60, 62, 63, 65, 69, 70, 71, 74, 77, 152, 153, 156, 158, 163
soil erosion 69, 70, 74
soil fauna 63, 64, 67, 77
soil flora 64, 64,
sources of food 15
sundaland 134, 136, 137

sustainability 1, 17, 28, 34, 36, 37, 42, 43, 44, 45, 46, 47, 48, 50, 53, 56, 59, 71, 72, 74, 147, 148, 149, 150, 151, 152, 153, 154, 155, 156, 157, 157, 158, 159, 160, 162, 163
  biodiversity-sustainability links 148, 149, 163
  sustainable development 43, 125, 137, 147, 150, 157, 150, 158, 159, 160, 162, 163
sustainable future 25, 37, 38, 53, 115, 150

T
threats to biodiversity 97, 107, 113
  agriculture 25, 25, 26, 27, 28, 29, 30, 31, 32, 34, 35, 36, 37, 38, 39, 40, 41, 42, 43, 44, 45, 46, 48, 49, 50, 53, 55, 56, 57, 58, 70, 72, 74, 100, 101, 108, 110, 120, 138, 139, 140, 155, 157, 159, 163
  environmental pollution 29, 101, 113, 150
  habitat fragmentation 98, 107, 110, 113
  hunting 100, 110, 113, 120
  overexploitation 100, 110
  poaching 100, 122, 128
  smuggling 100, 108
trophic dynamics 170
trophic level 39, 81, 105, 109, 111
tropical areas 8, 9, 170
troposphere 44, 153, 158, 170
tundra 170

U
UN Law 87, 92
uncultivated land 38, 42, 43, 45, 47, 48, 53, 54
UNEP 147, 163
UNESCO 125, 127, 128, 139, 142, 143
United Nations 14, 32, 33, 87, 147
unsustainability 30, 37, 52

V
values of biodiversity 13, 19, 23
  bequest value 17, 19, 21, 23
  cultural and aesthetic 16
  direct values 14, 19
  drugs and medicines 15
  existence value 14, 16, 19, 21, 116
  indirect values 14,
  option values 17
vent ecosystems 83, 86
vegetarianism 31
vermicompost 156, 159

W
water 10, 17, 18, 42, 43, 45, 46, 60, 61, 83, 84, 88, 89, 92, 99, 101, 102, 103, 106, 116, 124, 128, 130, 131, 139, 153, 155, 156
water cycle 170
WCED 147, 157, 159
Western ghats 11, 12, 19, 52, 68, 136, 137
wild relatives 15, 50, 52, 54, 57
wildlife 16, 32, 33, 53, 58, 101, 102, 108, 117, 120, 122, 123, 124, 132, 138, 139, 140, 142, 143
  man-wildlife conflicts 101, 102
  measures to avert man-wildlife conflicts 102
wildlife sanctuaries 16, 102, 117, 122, 123, 124, 138, 139, 140
WWF 32, 138, 140, 147

Z
zoo 132, 133, 134
zoomastigina flagellates 67
zooplankton 170